Pasture Management for Horses and Ponies

Pasture Management for Horses and Ponies

Gillian McCarthy

BSP PROFESSIONAL BOOKS

OXFORD LONDON EDINBURGH

BOSTON PALO ALTO MELBOURNE

First published by Collins Professional
 Books 1987
Reprinted by BSP Professional Books
 1988

British Library
Cataloguing in Publication Data
McCarthy, Gillian
Pasture management for horses and ponies.
1. Horses—Feeding and feeds   2. Grazing
—Management
I. Title
636.1'084   SF285.5

ISBN 0–632–02286–8

BSP Professional Books
A division of Blackwell Scientific
 Publications Ltd
Editorial Offices:
Osney Mead, Oxford OX2 0EL
 (Orders: Tel. 0865 240201)
8 John Street, London WC1N 2ES
23 Ainslie Place, Edinburgh EH3 6AJ
Three Cambridge Center, Suite 208, Cambridge
 MA 02142, USA
667 Lytton Avenue, Palo Alto, California
 94301, USA
107 Barry Street, Carlton, Victoria 3053,
 Australia

Printed and bound in Great Britain by
Mackays of Chatham, Kent

Contents

Foreword

The author of this book, though highly qualified in all aspects covering the careful management and development of the horse, has made a thorough study of pasture management. She has gone to great lengths to get valuable information from scientific sources which are important for the layman, who may spend endless hours, even lifetimes, improving and maintaining grassland and its nutritional output. This information is essential for breeding a sound and healthy individual, whether for the race track or the showring.

The subject of grassland management, which covers many aspects including quality of land, drainage, suitable grasses, herbage and treatment, could be the subject of many volumes. This is because the area of grassland on which the horse is left may form up to 100% of its daily environment. Therefore careful grassland management must be maintained to achieve the best possible results.

The subject matter is thoroughly sound and therefore I recommend this book to those with large or small establishments who wish to improve the standard of their pastures.

J. A. Cowe
Juddmonte Farms
Wargrave

Preface

Whether you call it a field, a paddock, a pasture or a corral, the area of grassland on which a horse is kept may form up to 100% of its daily environment, including providing up to 100% of its nutrition and up to 100% of its exercise. This may apply to the thoroughbred foal who is potentially worth millions, and equally to the Thelwellesque family pony who is beyond price to his particular family, and in between the whole range from 'meat money' and upwards.

Plate 1 Pasture plays an important role in the environment of the brood mare, and has done for a long time before George Stubbs (1724–1806) painted this picture of 'Pumpkin with Stable Lad', now the property of Mr and Mrs Paul Mellon. (*Photograph taken by Christopher Moore.*)

Precious little attention has been paid to the management of such a valuable resource specifically for horses, whether they be pets, working animals or high performance athletes. There is much confusion amongst horse keepers on how to manage to best advantage their half acre at the end of the garden or 500 acres of prime stud land on the chalk downlands or rolling acres of Kentucky.

The only way to manage anything effectively is to understand 'what makes it tick'. The object of this book is to impart the relevant background information, where possible applying it specifically to equine situations, whether for thoroughbreds, family ponies, donkeys or even zebra, to help the horse keeper to make reasoned decisions on how to manage his pastures and when and whom to ask for help.

Technical terms have been explained in the text or glossary, and practical advice has been given in reasonable depth. It might appear helpful to go further but a line has to be drawn at 'how to drive a tractor' and so forth. Perhaps 'whether or not you need to drive a tractor and if so what sort' will set people on the right road. There are plenty able to answer your practical questions once you know what questions to ask. I have, however, pointed out where common mistakes are made.

A field can be a pretty unfriendly environment to a horse who is confined (by man), what with wire and worms, poisonous weeds and polluted water, and sadly these occur not just in urban fringe 'horse slums' but even on all too many of the best known studs. It's time to take a good hard look at pasture management and put some more effort into both research and education, and I hope this work will provide an informative handbook for the private horse owner, a management aid for the 'professional horse manager', including studs, and a textbook for students.

There will undoubtedly prove to be better ways of doing some of the things I've suggested, but I hope this book will be the catalyst that sets some horse managers, researchers and even commercial organisations on to thinking about those better ways specifically for horses.

Gillian McCarthy

Acknowledgements

My thanks are due to all those who have assisted me in producing this book, and especially to my parents .for their monumental help and support throughout and for compiling the index, my sister Clare for battling with my temperamental word processor to type the manuscript in her 'spare time', Jane Holmes for deserting her own business to complete the manuscript and type the poisonous plant tables, and Pip Browning for 'holding the fort' at the office.

Thanks are also due to Mark Smallman for taking time off from producing his lovely portraits of horses and dogs to produce the excellent drawings at short notice, to Christopher Moore for printing most of the photographs and enhancing my more amateurish efforts, and Annie Dent for the lovely cover photograph.

The advice of Steve Staines of Wessex Soil and Land and Tony Clarke of Wayland Drainage was invaluable and thanks are also due to Howard Petch, Principal of Bishop Burton College of Agriculture, for reading the final manuscript. I should also like to thank my editor, Julian Grover, of Collins Professional and Technical Books for his help and advice throughout.

Acknowledgements are also due to those manufacturers who provided information about the products mentioned throughout the book. Those who kindly supplied photographs are acknowledged in the text.

Further Reading

A Festival of Dressage
Jane Kidd

Dressage
Kate Hamilton

*Educating Horses from
Birth to Riding*
Peter Jones

Equine Injury and Therapy
Mary Bromiley

*Foaling – Brood Mare and
Foal Management*
Ron and Val Males

Guide to Riding and Horse Care
Elaine Knox-Thompson and
Suzanne Dickens

Horses are Made to be Horses
Franz Mairinger

Long Distance Riding
Marcy Drummond

Riding Class
British Horse Society

The Athletic Horse
Carol Foster

The Complete Book of Horse Care
Tim Hawcroft

The Complete Book of Ponies
Lorna Howlett

The Equine Veterinary Manual
Tony Pavord and Rod Fisher

The Young Horse
Elaine Knox-Thompson and
Suzanne Dickens

*The Performance Horse:
Management, Care and Training*
Sarah Pilliner

The Systems of the Horse
Jeremy Houghton Brown and
Vincent Powell-Smith

Chapter 1
Introduction

The horse, *Equus caballus*, evolved as a forage eater. Whatever lifestyle man and science attempt to impose upon the horse, they are unlikely to alter significantly the digestive capabilities imposed upon the equine by evolution, and that evolution has been dictated in the long term by environment. It is true that other domestic livestock have been bred to adapt to a new environment by man – there are pigs and poultry that would not last a matter of days out of doors on a 'natural' diet – but, so far, man has not selectively bred horses for growth rates and feed efficiency, he has bred for speed and size. Even horse meat consuming countries have only attempted to do limited work in this area, tending to 'retire' horses and ponies bred for other purposes or 'mop up' indiscriminate overproduction, rather than breeding for meat production.

Grassland is a seriously underutilised resource in horse management, either because horse owners have no idea how to manage their grassland or have attempted to manage it as for cattle and run into serious problems as a result, or because they simply can't be bothered or are unaware of what they could achieve with a given area of land.

Not all grassland related to horses is for grazing or conservation of hay or silage. The management of gallops and exercise paddocks, and the varieties of grasses in them to produce a springy turf which will not overstress valuable limbs, are quite distinct from the requirements for grazing and conservation, and are beyond the scope of this book.

Very little work has been undertaken on the subject of grassland management specifically for horses. The State Extension Services in the USA provide a service of variable quality, although as some states have specific equine specialists they are able to provide suitable advice. In the UK, the Ministry of Agriculture, Fisheries and Food's Agricultural Development and Advisory Service (ADAS) does not officially cover horses (which are not classed as 'agricultural animals'), although they do offer a basic soil analysis service. One worker who has conducted specific studies is Mrs Mary Tavy Archer, formerly of the Equine Research Station at Newmarket, whose work is mentioned below.

Different commercial interests are also able to offer advice, of variable quality. Some merely transpose their knowledge of cattle and sheep pastures for horses, which is *not* likely to be very successful, whilst others do attempt to build up specialist knowledge on the subject. These companies include some of the herbage seed houses and fertilizer, trace element and agrochemical companies. As far as I know, Mrs Archer and I are the only independent consultants in the UK advising on all aspects of pasture management for horses. Some larger and more prestigious studs (breeding farms) in the USA have their own agronomists to advise on both their pasture management and management of crops for conservation from outside producers – laying down ground rules for fertilizer treatment, etc. Sadly, very few stud owners in the UK even know what the word agronomist means but muddle through using advice largely developed for cattle pastures, or having run into problems using that system, barely 'manage' their pastures at all because they have been frightened off by overuse of fertilizers, etc. Pasture management is a matter for balance, as is nutrition, and the two subjects are indivisibly interwoven in horse husbandry.

To the stud or breeding farm manager, the need for good grassland management should be self-evident. Pasture provides invaluable nourishment and exercise for all breeding and youngstock. The better managed the grassland, the better the start in life for youngstock, the better the quality and quantity of lactation of brood mares, and improved conception rates are achieved by healthier mares and stallions. The importance in this context, for example, of beta carotene, an important constituent of fresh grass, and to a lesser extent conserved forages, should not be underestimated, and likewise folic acid, for prevention of anaemia in the mare.

For the ordinary horse owner, the advantages or even possibilities of good pasture management are less self-evident. Having said this, the continuance of 'horsiculture' on urban equine slums is inexcusable. Keeping horses in bare, weed- and worm-ridden 'paddocks' with barbed wire or metal fence posts, corrugated iron sheets, etc., to satisfy a human desire to keep horses, however much you 'love' them, is simply not good enough (plate 2). There are 'accidental accidents' and 'inevitable (and, therefore, avoidable) accidents'. Horses, barbed wire, metal fence posts and corrugated iron simply do not mix. If paddocks are an eyesore and look messy and unkempt (however much work the owner may expend on them), they attract vandals, upset non-horsey local people, and are irresponsible. Sadly, you *do* need money in order to afford to keep horses 'in the style to which they ought to be accustomed' and if you don't have it, or the time and know-how required, however much you desperately

Plate 2 This bleak, weed-infested paddock with dangerous mesh fencing and 'greener grass' on the other side is reflected in the poor condition of the pony.

want a horse or pony, it is not fair on the animal unless and until you can afford it. Life is unfair to people, but people can be desperately unfair to horses. However, for the amateur horse owner who does have, or can obtain, reasonable facilities, a great deal of pleasure, as well as a significant contribution towards a horse's keep and exercise, may be obtained through sound pasture management.

Horses *can* be worked fit off grass. Many endurance horses, working at the peak of athletic performance have proved this time and time again. For the performance horse in general, however, time at grass usually means an hour or so a day for a spot of self-exercise and a change of air, or being turned out and virtually forgotten ('roughed off') for a 'holiday' out of season. For an athlete, this is about the worst thing that can happen, but management of the equine athlete still owes much to the whim and 'instinct' of the horseman and relatively little to our growing scientific awareness of the appropriate way to manage it, both for performance and for its own lifetime psychological and physical fitness and welfare. It should be remembered that if a performance horse is going to go 'out to grass' for more than just a 'stretch of the legs and a spot of sunshine' the better the quality of that grass – the better its nutrition, the sooner it will come back fit – and the better the physical make-up of the sward, the safer will be the equine athlete's precious limbs.

Apart from breeding stock and the athletic performer, we have three other basic types of horse at grass: the working horse, e.g. from a riding school, the livery 'farm' horse and the family pony or 'backyard horse' (USA).

Special considerations

The working horse

Some riding schools, showing yards, etc., have very definite grassland management policies. They may obtain significant amounts of keep from grass, plus forage for conservation and exercise or 'loafing' paddocks. They may or may not use other livestock as part of their management system. The pasture is important to them economically. They are likely to own it or be long-term tenants who are able, depending on money and labour availability, to have a long-term plan of management and investment in the land. The more reputable establishments will also have a desire to maintain certain standards of appearance and keep on the right side of planning authorities. Unfortunately, some are too ignorant or inefficient to do so. This way the horses may suffer to a degree, or at the least more bought in feed will be required. Also, with a growing awareness of the 'environment', unsightly weed and broken-jump infested fields, with motley lean-to shelters, railway waggons and broken fencing, upset local inhabitants, planning authorities and animal welfare workers, and by doing so make it harder for other horse owners as local authorities clamp down on 'horsiculture' (*sic*). More of this later.

The livery farm

Livery farms vary widely in standard. The horse owners usually have little control over the grassland management, other than by withdrawing their custom, which may not be practical in areas of limited resources. In an ideal world, horse owners would be prepared to pay more for well-managed, adequately fenced horse paddocks, than 'horse slums'. In a limited market they take whatever they can get, often paying through the nose for it. In urban areas there is often little incentive for livery (boarding) farm owners to expend time and effort on grassland management, when they can extract a comfortable living from doing little more than owning or renting the land and putting a sign up on the gate. There may even be a strong disincentive.

Studies of 'horsiculture' in urban areas have shown that renting of fields to horse owners on a short-term basis can be used as a long-term technique to get round planning and 'green belt' restrictions. The scenario is something like this: the property developer buys a tract of agricultural land on the edge of an urban area with no hope of obtaining planning permission for building. He divides it up and lets it out to land-hungry horse owners on short lets. He does not attempt to improve or even maintain the quality of the grazing, and is likely to intentionally overstock, both for immediate income and for satisfaction of his long-term ends. Because they only have short-term rental agreements, and many different owners keep their ponies on the same fields, most cannot agree to a group expenditure of effort or money on grassland management, or the risk of being turned off the field as soon as they have done the work. So, the land deteriorates and becomes 'horse sick', scrubby weeds appear, grotty old baths are used as water troughs and unsightly (cheap) sheds proliferate. After five to twenty years the local planners become so disgusted with the eyesores they *persuade* or *allow* the *poor* unfortunate land owner to blossom into a fully-fledged property developer (reluctantly of course) and sell his land at a full profit for a housing estate, where pony-mad children move in and desperately want a pony, and grazing for it, so the *poor* man has to buy some *more* land and start all over again!

It is not unknown for local authorities to follow this practice with their own land. Happily, there are some livery farms which keep their paddocks safe and productive, and do not overstock so that fields can be rested and do not become 'horse sick' or at least practise 'rotational grazing'. These should be cherished and are worth paying for. The owner who overstocks to gain more income, or because he is too soft-hearted to say 'no' to a desperate horse owner, cannot rest paddocks, or even fertilize them as he cannot keep stock off whilst the fertilizer breaks down. With twenty to thirty horses on five acres (not unknown in urban areas) he certainly won't be picking up droppings and will find harrowing to spread them pretty tricky if he can't take the stock out of the field.

The biggest problem with keeping horses at grass livery, good or bad, is that unless the landowner imposes a contract whereby *all* animals are wormed on arrival, and then at regular intervals at the same time, worm control becomes a virtual impossibility. Livery farm owners should be encouraged to follow this practice. The horses on their farms will be healthier, use feed more efficiently, and they will be able to buy wormers in bulk, and even make some profit on them.

The 'pony-at-the-bottom-of-the-garden' or 'backyard horse'

Many private horse owners, and particularly pony owners, have a non-rural background combined with a fear of lush grass which they probably equate with 'managed' grass. The farmer up the road, whose children's ponies contracted laminitis (fever-of-the-feet, founder) when he put them out with his dairy cows has probably told them 'you can't use fertilizer on horse paddocks'. Either that, or he has stopped making hay on his farm and so no longer has the equipment to 'pop along' and cut or 'top' their extra grass if they have any, and neither do local contractors if most of the farmers in the area have gone over to silage. So grass that actually *grows* is regarded with the sort of distrust science fiction affords to 'the triffids' (homicidal plants).

For ponies, the 'starvation paddock' system is often intentionally favoured. The paddock is kept bare except for a colourful if untidy selection of weeds, some of which are almost illegal 'notifiable weeds', such as certain docks and thistles which spread like wild fire. However, these are generally ignored by the authorities until an irate neighbouring farmer complains about the weed reservoir over the hedge continually re-infesting his carefully managed arable land, or somebody complains about the eyesore, or until local authorities get more worked up about so-called 'horsiculture' and start looking for any excuse to impose planning restrictions on horse keepers. (It has to be said that local authority owned land is often one of the worst offenders in this 'weed reservoir' effect.)

Other weeds may contain toxic substances but may not *normally* be eaten by livestock. However, they may well be palatable when wilted, which can happen if they are damaged by galloping hooves. On a starvation paddock, a desperate pony tired of eating fences and barking trees in its craving for fibre and minerals may well set to on the potentially poisonous bracken, buttercups (yes, pretty little buttercups are poison-ous) and worst of all ragwort (plate 3), and even acquire a taste for them. This I suspect may well explain, along with vitamin and mineral deficiencies, the sluggishness of many children's ponies – a subclinical slow poisoning from the pasture. Even apparently harmless but non-palatable weeds are robbing the soil and edible herbage, and ultimately the pony, of vital nutrients, especially trace elements.

It seems likely that 'starvation' is most certainly not the best policy for laminitis-prone ponies, causing further hormonal imbalances and a greater 'over-reaction' when spring grass or any other available carbohydrate-rich feed source is introduced, and a calcium deficiency may further exacerbate this effect. Better perhaps to keep the pony supplied with low-quality roughage such as straw or molassed chaff, plus

Plate 3 Ragwort (*Senecio spp.*) is a common poisonous plant in horse pastures which must be eradicated before the horse can graze safely. It is particularly palatable once wilted or in hay.

a feed block or pony nuts to supply essential minerals, vitamins and protein, and always a trace mineralised salt lick.

The pony stud

Studs breeding native ponies and children's ponies will need to consider that the native pony evolved to be very hardy and a highly efficient utiliser of feed, giving it the ability to gain maximum nourishment from poor quality fodder. Breeding a pony of this type off the moors, mountains and forests, on a lusher, lowland farm will not magically reduce its feed utilisation efficiency. This would take thousands of years of evolution, or at the least hundreds of years of careful selection and breeding for an 'inefficiency trait'. So, the stud owner should not be surprised that when keeping such ponies in what is to them an 'oversupply situation', however many generations ago they were taken from their native habitat, they become obese and come down with the typical 'show pony' diseases of hyperlipaemia and laminitis, plus various fertility and behavioural problems.

Thus, the pasture on a pony stud must be carefully managed. Grass varieties of lower feed value may be introduced, clover most definitely eradicated, and a tight grazing/topping/mopping up with sheep/cattle policy developed. Fertilizers should be used judiciously and attention paid to liming, and as the grass will be strictly controlled, weeds should be kept down. As native pony breeding stock and youngstock may receive minimal supplementary feeding, trace-mineralised salt licks are essential, herbs may be introduced into the pasture, and the use of feed blocks may be considered.

The thoroughbred stud

The thoroughbred stud is a 'whole different ball game', particularly if the aim is to produce stock destined to race at two years old (plate 4).

Plate 4 Superbly managed pasture at a top stud in the UK. Note the top rails of the fences are flush with the tops of the posts, with double fencing and gates between paddocks.

Youngstock must grow on rapidly but in a balanced way and to this end they, and also the gestating and lactating mare, will require optimum amounts of balanced nourishment, a fair proportion of which may be obtained from the pasture. A further consideration is safety of valuable youngstock, whose limbs must be protected from injury by development of a good 'bottom' to the turf.

The 'upbringing' of a young thoroughbred, i.e. its environment in relation to nutrition, stress exposure, exercise and handling, disease and parasite control, plays an important role in producing a potentially high class athlete. A horse cannot race by genetics alone.

The moral question

What on earth does morality have to do with grassland management? You may well ask. There is a field near my home where somebody has written on the gate:

'Feed humans, not horses'

At a time when a huge proportion of the people on this planet do not have enough to eat, and as the gap between 'north and south' widens, the wasteful behaviour of affluent nations whilst others starve, rather than specific political doctrines and flexing of muscles by the superpowers, could well be the 'straw that breaks the camel's back' and the trigger that fires the starting gun for World War III. Hysterical hyperbole? Perhaps. The Independent Commission on International Development Issues whose 'North – South: A Programme for Survival', also known as the Brandt Report, which was published in 1980, do not think so.

What's all this got to do with your two acres at the bottom of the garden? Well, let's not get this out of proportion. A nation's health depends not only on food and environment, but on leisure and exercise. The last two may not be very prominent in the minds of the poorer southern hemisphere, and as food shortages seem to be partially a problem of distribution rather than total world production, it would seem unreasonable to expect the richer northern hemisphere to sacrifice its leisure pursuits, especially as leisure time is on the increase as the ubiquitous micro-chip makes its mark.

Your weed-infested two acres could probably support several Ethiopian families. Impractical? All right, 'what they can't see won't hurt them'. Political activists *do* see though.

Improving knowledge of ponies at pasture *could*, by diffusion of know-how, help the Third World countries that employ horses, mules and donkeys as a major part of their rural economy. Examples of this are already happening. The Donkey Sanctuary at Sidmouth in Devon, England, currently has some two thousand donkeys on its books. These are intensively managed on a number of farms. Some are used for work, for example at the Slade Centre – their Riding for the Disabled Centre, and their enviably well-equipped Donkey Hospital must lead the world in research into diseases and management of the donkey. What's the point? How does that help the Third World? Well, to me, the existence of this super-efficient organisation is justified by the fact that knowledge gained in the UK operation is applied by a third charity, run from the same centre, called the International Donkey Protection Trust. This operates far-reaching projects in the Third World which not only improve the welfare of the donkeys but, by making them healthier and live longer, improve the prosperity and standard of living of their 'peasant' owners.

So, a useful motto for the modern horse keeper might be

'Don't let horses waste resources'!

Specific management systems, seed mixtures and fertilizer practices will vary considerably in different parts of the world, and even within a country as tiny as the UK, with its variations in terrain, climate and soil type. Thus this book aims to provide a discussion of general principles with specific examples, to help the horse keeper make reasoned and balanced decisions in formulating a pasture management policy, and tell him or her where to find help if needed from the lists of useful addresses and publications provided at the end of the book.

Chapter 2
Grazing Paddocks – physical requirements

Many factors affect the suitability and treatment of grazing land for horses: altitude, aspect, rainfall, wind speed, direction and temperature, soil type and drainage.

The first three cannot be influenced practically by man. The effect of wind can be modified by using stout hedges, shelter belts of trees and building suitably sited field shelters. Soil type may possibly be modified in future by the use of new materials which, when incorporated in the soil, can bind sandy soils together whilst improving the drainage of clay soils. These are likely to be expensive, but may be worthwhile for use around gateways and water troughs. The organic matter content of soils may also be increased by addition of organic materials. Drainage can be affected significantly by man in most instances and is the first subject for attention when renovation of paddocks is undertaken. A good stout hedge in the path of the prevailing wind can do much to shelter grazing animals and, by reducing their feed requirement to keep warm, may enable them to maintain body 'condition' with less supplementary feeding than those in unsheltered paddocks. Horses do not mind the cold, but a cold wind, especially if the horse is wet, is quite another matter. Man-made fencing affords little shelter, although some new wind breaks are being tried which involve strips of PVC fixed on a frame hanging perpendicularly to break up the wind.

Soils and horse pasture management

A full appreciation of soil is of vital importance when considering pasture management. The nature of the soil and its interaction with the weather determine not only its ability to grow grass, but also the conditions underfoot and the quality of that grass which, in turn, controls the supply of nutrients to animals under a grazing regime. Because soils vary, so does their suitability for horse pasture in the same way that not all soils are suitable for arable cropping. An understanding of soil, its problems and its pitfalls, and a knowledge of ways of improving and managing it are undoubtedly necessary for successful pasture management.

Information about soil characteristics and distribution derives mainly from the relevant soil survey organisations. In the UK these are the Soil Survey of England and Wales and the Soil Survey of Scotland. These organisations map and study soils and they have produced maps at varying scales for most of the country. In lowland England and Wales about a quarter of the country has detailed soil maps suitable for interpretation at field level. The rest of the country is covered by reconnaissance soil maps and reports that are useful only in a general sense. Soil investigations can be commissioned from the above organisations as well as from independent consultants.

Soil type

Soils are very complex. There are some 850 soil types, or 'soil series' in England and Wales. However, some basic principles do apply to all soils and an appreciation of soil–plant–climate interactions is easy to obtain.

Soil itself is made up of mineral particles, organic matter (humus) and decaying plant and animal debris, air and water. Soil fertility is defined by the basic physical characteristics of the soil: texture, depth, height of the water table, drainage characteristics, chemical composition of the soil and parent rock and organic matter content. Soils are all mixtures of particles – the predominant particle determines the name of the soil (table 1).

Table 1 Soil type

Particle type	Clay (%)	'Blow-away' sand (%)	Loam (medium) (%)
Coarse sand	9	49	20
Fine sand	10	44	43
Silt	21	1	14
Clay	52	2	15
(Humus)	(8)	(4)	(8)

Soils are built up over bedrock, which is covered by subsoil from half to several metres deep, which is topped by topsoil from a few centimetres to about one metre deep in some situations. Loam soils range from sandy through medium to clay loams and are probably the commonest soil types in temperate regions. A sandy or sandy loam soil is regarded as 'light' soil and clay or clay-loam as 'heavy' soil. Permanent grass tends to thrive on clay soils and is often the preferred crop for this soil type as it can become waterlogged and difficult for heavy arable crop machinery to operate on in winter.

Different particles arise from different rock sources so different nutrients are present in them. Clay soils hold more water and are inherently well supplied with nutrients. Unlike clays, sandy soils have no *colloidal* attraction for nutrients, and as water washes through easily, nutrients are quickly leached (washed) out. Clay soils tend to be poorly supplied with oxygen, especially when waterlogged. The pH or acidity of the soil is also important for optimum grass growth. Soils which are too alkaline (non-acidic) (e.g. some chalky or calcareous soils) tend to 'lock up' minerals so that they are unavailable to plants and the animals eating them. Sandy soils which drain easily lose calcium easily (by the action of weak carbonic acid in rain). As calcium corrects acidity, this usually means that sandy soils tend to be acidic.

Most crops have a preferred level of acidity for growth. In the case of temperate grasses this is a pH of 6.5 to 6.8 (the pH being a logarithmic scale used to express acidity), which can be easily checked by your local Agricultural Advisory Service or fertilizer company. A rough estimate can be obtained using a low-priced kit available from horticultural suppliers and garden centres. The more expensive 'dial monitors' are of more dubious accuracy at present.

As a general rule liming (addition of ground calcium carbonate – limestone) to add calcium and correct acidity is likely to be necessary every four years for sandy soils and every seven years for clay soils. Rarely, even some calcareous ('chalky') soils may require liming as the calcium in them can be 'locked up' and unavailable. Many of the best horse breeding regions of the world are on chalky soils, but it is not safe to assume that all these soils are of a suitable pH for optimum growth of pasture and horses.

A further soil type is the organic or 'peaty' soil, which tends to be low in mineral content and high in organic matter with a tendency to be acidic. It is formed in wet conditions and tends to hold water like a sponge, so drainage can be very difficult. As a rule, drier sandy soils and sandy loams tend to dry out sooner in spring and be 'earlier' whilst grass on clay soils may be slow getting started in spring, but will tend to be more drought resistant. Thus a stud or livery farm with a variety of soil types can make a reasoned choice of which paddocks to use first in spring and last in autumn.

Soil identification

Soils are usually layered and it is the nature of these layers that determine soil characteristics. Soils are recognised according to their texture, depth, natural drainage, some gross chemical characteristics such as calcium carbonate (or lime) content, and the nature of the parent material from which the soils have developed. Of these features texture and drainage characteristics exert most influence on land use.

Soil texture is the proportion of sand, silt and clay that the soil contains and soil textures are given names as illustrated in table 2. Where the soils contain appreciable amounts of organic matter as well, they are described as humose or peaty according to the amount of organic matter. Texture is important in determining soil qualities. For instance, sandy soils hold little moisture against periods of drought and they are often droughty. Clayey soils often have dense, slowly permeable subsoils leading to waterlogging and wet conditions during the autumn, winter and spring with subsequent structural damage and poor crop growth. Use of terms like sandy soils and clayey soils can be misleading in some senses, especially if these refer to the topsoil texture alone. The qualities of the subsoil are just as important and modern soil description takes this into account.

Texture describes the mixture of different particle sizes in soils and names such as sandy loam and clay loam are used to describe these mixtures. Coarse-textured soils are also referred to as 'light' soils to indicate ease of cultivation and conversely the term 'heavy' is used to describe clays.

Classification of texture

The Agricultural Development and Advisory Service (ADAS) and the Soil Survey for England and Wales (SSEW) use the same system of classification. This contains eleven major classes based on the proportions of three size fractions:

Clay less than 0.002 mm diameter
Silt 0.002–0.06 mm
Sand 0.06–2.0 mm   (fine 0.06–0.20 mm)
                     (medium 0.20–0.60 mm)
                     (coarse 0.60–2.00 mm)

The eleven classes are defined in fig. 1. Sand, loamy sand, sandy silt loam and sandy clay loam classes may be subdivided according to the size of the sand fraction:

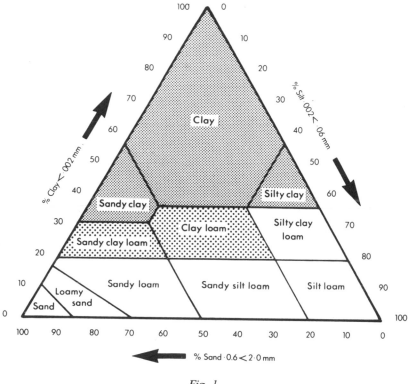

Fig. 1.

Fine – more than two-thirds of sand less than 0.2 mm.

Coarse – more than one-third of sand greater than 0.6 mm.

'No prefix' – less than two-thirds fine sand and less than one-third coarse.

Soils formed on chalk and limestone usually contain natural calcium carbonate (lime) and are denoted by the prefix 'calc'. For many purposes, individual classes with similar properties are grouped, as in table 2.

Identification by hand texturing

Usually a particle size distribution (psd) is not available and texture has to be assessed by hand texturing, that is, from the feel of moist soil (fig. 2). For most soils there is a close relationship between psd and hand texture.

Take about a dessertspoonful of soil. If dry, wet up gradually, kneading thoroughly between finger and thumb until aggregates are broken down. Enough moisture is needed to hold the soil together and for the soil to exhibit its maximum cohesion.

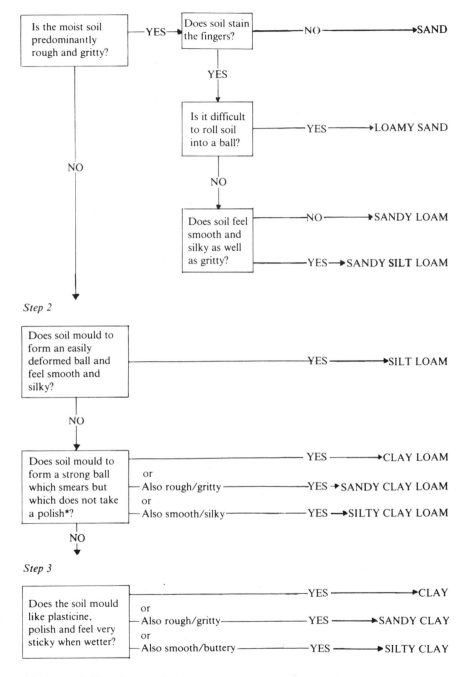

Step 1

Step 2

Step 3

* Polish: smooth shiny surfaces to soil when rubbed

Fig. 2 Identification of soil by hand texturing.

Table 2 Texture groups

Texture group	Texture class	Abbreviation
Sands	Sand	S
	Loamy sand	LS
Light loams	Sandy loam	SL
	Sandy silt loam	SZL
Light silts	Silt loam	ZL
Medium loams	Sandy clay loam	SCL
	Clay loam	CL
Medium silts	Silty clay loam	ZCL
Clays	Sandy clay	SC
	Clay	C
	Silty clay	ZC

Natural drainage is very important as waterlogged soils usually have many problems. Soil wetness results from a high groundwater table, or a slowly permeable subsoil, or a combination of both. Soils affected by high groundwater usually occur in low lying places. Natural drainage can be supplemented by laying pipes or other methods of artificial drainage, but it should be noted that with the exception of very permeable soils affected by a high water table, artificial drainage will rarely make a poorly drained soil well drained. The soil can be improved but not necessarily transformed.

For artificial drainage to be most beneficial it is necessary to identify the source of the soil wetness so as to allow the correct drainage system to be employed. Permeable soils with high groundwater are usually drained by placing either plastic pipes or clay tile drains at about one metre (approximately three feet) depth and connecting them in such a fashion as to lower the groundwater table and remove excess water to ditches or in extreme cases to pumps. Such drainage usually produces very satisfactory results with a marked improvement in soil conditions. Most of the Fenland in East Anglia has been drained in this fashion.

Slowly permeable soils will have inherent soil wetness which results from rainfall 'ponding' in the soil during the autumn, winter and spring. This usually occurs in clayey soils and the most common remedial technique is a combination of pipes covered with a gravel permeable fill linked by 'mole' drains. There will be some soils, however, that have dense, slowly permeable subsoils but are not suitable for moling. Many soils have loamy upper layers over clayey subsoils. Moles drawn through this loamy material do not last more than two or three years whilst moles drawn through suitably clayey soil can last up to ten years. Hence an appreciation of soil type is essential in drainage design (see below).

Categories of soils

The soils of Britain fall into a small number of broad categories which are based on major soil characteristics.

Shallow soils, less than about 40 cm deep, occur over hard rocks like slate in the West Country and Wales, and also over limestones and chalk in lowland England. Shallow soils over rocks like slate are droughty and acid in contrast to shallow soils called *rendzinas*, which occur over limestone and chalk. Such soils are usually calcareous and are traditionally thought of as giving some of the best horse pasture. Soils over limestone, in low rainfall districts, are usually quite drought prone but those over chalk are less droughty because plants can extract water from the chalk which acts as a sponge. All of these shallow soils are well drained, suffer minimal wetness and can be trafficked for much of the year with little sward damage. These rendzinas often carry springy natural grassland with a wide range of herb species where they are unfarmed.

Deeper well-drained soils are *brown soils* which have brown, evenly coloured subsoils. They are very common over a wide range of deposits in England and Wales and the lowlands of Scotland. They are loamy, sandy or occasionally clayey. Depending on the parent rock, i.e. whether it is lime rich or not, they can be calcareous or acid. In general, where these soils are relatively deep and not too sandy, they have good reserves of soil moisture and carry livestock well with little damage from poaching. Some sandy brown soils can be very droughty, however, while some of the clayey brown soils can have a restricted grazing season, especially in the wetter parts of the country.

Soils which suffer periodic waterlogging are called *gley soils*. They have grey, mottled subsoils with rusty colours which indicate prolonged subsoil waterlogging. Those that are permeable and have a high groundwater table are called *groundwater gley* soils and those that suffer waterlogging because of a dense, slowly permeable subsoil are called *surface water gleys*. Groundwater gleys occur principally along river valleys in alluvium and also in low-lying places like the Fens. Surface water gley soils are very widespread in lowland England and occur on a variety of clay formations and also on boulder clays. They are the most extensive soil type in the lowlands and even after artificial drainage, can have a shortened grazing season and low potential productivity.

Upland moors and lowland heaths usually have very leached, very acid, nutrient-poor soils called *podzols*. On lowland heaths these are usually very droughty sandy soils although some can also be waterlogged due to the presence of groundwater.

Peat soils are found where conditions were originally so wet that plant remains have not decayed, but built up into thick layers of organic matter or peat. In the natural state these soils are very wet and ground conditions are boggy. In places like the Fens, reclamation and drainage have allowed these soils to become the most productive in the country, but uplands like the Pennines and Wales have extensive acid peat bogs which are unsuited to farming.

Table 3 Available water reserves

Soil texture	Percentage by volume (mm/10 cm depth)	
	Topsoil	Subsoil
Sand	8	5
Loamy sand	12	9
Sandy loam	17	15
Sandy silt loam	19	17
Silt loam	22	21
Clay loam	18	15
Sandy clay loam	17	15
Silty clay loam	18	15
Clay	18	15
Sandy clay	17	15
Silty clay	18	15
Peats	35	35

Table 3 contains average figures for available water reserves from a range of values for each texture. The range depends on variation in particle size, organic matter and compaction. Approximate available water reserves for a soil profile were calculated by aggregating values by depth for the appropriate textures.

A good depth, say 300 mm (12 in) of workable soil is desirable for good root development. Peat and sand soils may be much deeper, whereas shallow soils may occur over pure clay or over rock (e.g. UK Wolds).

For the larger horse enterprise, particularly for example when choosing a location for a new stud, or establishing rents for a stud farm, it may well be prudent and extremely useful to have a full soil survey done of the land in question, and to check whether the land is covered by existing soil survey maps. This exercise can also provide valuable information for predicting accurately the going on gallops and race courses.

Contact the Soil Survey of England and Wales, or soil surveyors listed in the membership list of the British Institute of Agricultural Consultants, both addresses being given at the end of this book.

Soil suitability for horse pasture

Nutrition

Assuming that a large proportion of the nutrition of the animal comes from grass and the grazing season needs to be as long as possible, then some soils are more suitable for horse pasture than others.

In order for grass to grow, it needs nutrients and water. Nutrients come either from the soil itself, or from applied fertilizer. Clover supplies nitrogen to the soil via nitrogen-fixing bacteria which occur in close association with the clover roots. Since a large proportion of clover in the sward may be inimical to horse health, nitrogen, which is essential for grass growth, may therefore have to be supplied in fertilizers, either organic or inorganic. Potassium and phosphate, the other two major plant nutrients, can vary markedly according to soil type and past management. Similarly soil pH is very dependent on soil type and if acidity is suspected a check on lime requirements is needed. Some soils lose lime faster than others and sandy soils, for example, require liming at more frequent intervals than clayey or loamy soils. In the same way sandy soils can lose nitrogen at a higher rate than other, heavier soils and more frequent applications of nitrogenous fertilizers may be required. Some soils fix phosphates very rapidly and annual doses may be required.

The best check on nutrient status is via regular soil testing by a competent authority who will also advise on rates of nutrients required. It must be borne in mind that grassland rates, especially of nitrogen, are usually recommended with high grass productivity for dairy cows in mind. Trace element disorders may be associated with some soil types but supplementation via feedstuffs can be implemented to reduce the risk of trace element problems.

Moisture

The climate of the UK is characterised by wet winters with an excess of rain and rather drier summers when transpiration by plants and evaporation exceed rainfall. In such a situation, plants have to draw upon the moisture supplies held by the soil. If the soil dries out, a soil moisture deficit develops. As different soils can hold different amounts of water it follows that in any given climatic situation, some will be more drought prone than others. The proneness of soil to drought is easily estimated from a consideration of the depth, structure and texture of the soil balanced against rainfall and potential transpiration.

The results of drought are clear – a cessation of growth and possible death of plants. The long-term consequences include the loss of a good sward, invasion by undesirable species of grass weeds better able to withstand drought, and bare patches which may be colonised later by mosses. The problem is best avoided by steering clear of very droughty soils in the first place. However, if this option is not available, irrigation may prove a useful alternative. Fixed irrigation systems are easy to run once installed and installation around the edges of paddocks can provide a problem-free solution. If your horse pasture burns badly in the summer irrigation should be considered (see below).

Apart from drought the other climate-induced problem is that of soil wetness. Poaching and damage to soil structure, a reduced grazing season and all the problems associated with wet ground conditions are major difficulties in UK farming. Although as mentioned above drainage can help alleviate the wetness, grass utililisation of paddocks can still be affected markedly even after drainage. It makes sense to check on soil type before starting on an enterprise. As well drained and poorly drained soils are often in close proximity, it is worthwhile having a soil map of the area involved. Paddocks can then be laid out so that, as far as is practicable, each paddock contains a uniform soil type. This allows sensible management, those paddocks with a shortened grazing season because of poor ground conditions being rested for longer than the better drained ground. It makes sense to conserve the soil structure and condition of your land.

If some structural damage has occured due to trampling or overstocking in poor conditions, much can be done to alleviate associated problems. The net result of trampling is compaction and this can lead to the immigration of unwanted species of herbage and can also cause some surface ponding in wet spells. Subsoiling using equipment like a 'paraplow' (see below) followed by rolling can help aerate the ground and improve structure. Slot seeding can then allow the appropriate sward to grow on the affected areas.

Maintenance of good soil structure is dependent on sensible management and control of stocking densities. Horses can be particularly hard on soil structures. Soil structures are naturally good under pasture. It is only overstocking which damages structure. In general the additives which are advertised as promoting good soil structure are expensive ways of buying lime. The answer to good soil structure is good soil management and not 'magic' material out of a bag. The rule is to keep horses off land when they do damage and to rotate them around paddocks so the soil has a chance to recover. The length of these rotations and the necessary recovery period is very dependent on soil types.

To sum up, different soils have different qualities and if an enterprise is being contemplated, then get the soil looked at first so that sensible decisions can be made. Similarly, if you have soil related problems, especially physical ones, do not expect miracles; the answer may lie more in your soil management than in a wonder material out of a bag. Soil management means suiting land use to the soil type.

Drainage

There is limited advantage in spending time and money on upgrading or renewing pasture in a field if the drainage is inadequate.

Drainage restores the air/water balance in the soil which is important for the development of roots and the soil organisms which break down organic matter, thereby releasing nutrients into the soil and improving soil structure. Well-drained soil warms up sooner giving earlier grass in spring and also makes the soil firmer, reducing damage in the autumn – so horses can stay out longer. Plants respond to good drainage by producing larger root systems, enabling them to draw on nutrients deeper in the soil and to be more drought resistant. Water running through the soil also prevents the build up of toxic substances (spray chemicals, etc.).

Drainage requirements will depend on soil type, rainfall and soil profile. Indications of poor drainage include irregular colouration (i.e. black, mottling, yellow/brown, green/blue, etc.) and generally grey subsoil colouring, and crops may look yellow (chlorotic) and unhealthy on waterlogged soil. They may also have blue edging on their leaves. Seed will be slow to germinate because the soil is cold and lacks oxygen; the surface may actually be waterlogged for all or part of the time; snow tends to linger in wet (or shaded) areas in spring and water loving (hydrophilic) plants, many of which are poisonous, will be established in bad cases. Land in valleys, or flat, low lying land, especially if below sea level (e.g. the Fens in the UK and Holland), or areas of shallow soil on very heavy clay, or light soils which have a 'pan' (an iron pan is usually a blue layer in the soil due to poor drainage; a plough pan is most likely on short-term grass leys where cultivations have been persistently at one depth causing smearing and sealing of the soil), or which have been compacted, are all liable to have drainage problems. (In 1972, out of the 11.25 million cultivatable hectares – 28 million acres – in Great Britain, some 5.6 million were below optimum production largely due to drainage problems.)

For the professional horse enterprise, contact your local government agency (in the USA the State Extension Service, in the UK ADAS) for advice about drainage and possible grants, or a drainage consultant who is a member of the British Institute of Agricultural Consultants, whose address appears at the end of this book.

Correcting drainage problems

Correcting drainage problems is largely a job for a specialist contractor, although mole drainage and ditch maintenance may be carried out by the horse owner. In the UK grants *may* be available when other livestock also use the fields. Contact your local MAFF/ADAS officer for information, although the grants are not as freely available now as in the post-war years. Some monumentally complicated form filling is also involved, as well as a possible means test type examination.

On established pastures:

(a) Check ditches for blockages, stagnant water, fallen banks, etc., and for silt and blocked drain exits. Ditches are like a river system – ditch runs into ditch, so you may need to check out ditches on a neighbour's land. Ditches may be all that are required to drain a field, especially on upland and hill farms. When cleaning the ditch out, keep it as straight as possible – on bends the current slows down and silting and eventual blockage occur. Keep an even gradient slope – steps lead to silting. Keep ditches fenced off to avoid damage to ditch sides and potentially fatal accidents. Be especially sure that small foals cannot roll under the fence and drown in the ditch. I have come across an alarming and entirely wasteful number of tragic cases of this type. Keep to the correct depth, which depends on the outlets, don't make the ditch deeper. Keep the shape correct. Each soil type has a 'natural shape of retention', clay soils have the steepest sides. Do not dump spoil next to the ditch, spread it away – otherwise the weight may cave in walls or at the least a good proportion may be washed back down to silt up the ditch! Figure 3 illustrates problems related to silting.
(b) Maintain culverts (i.e. the pipes taking the ditch under gate openings, etc.) in good condition. Normally a culvert pipe is at least 225 mm (9 in) diameter.
(c) Check the junction where the ditch joins a piped main, clean out the silt-trap if there is one.

(a)   Problems relating to silting

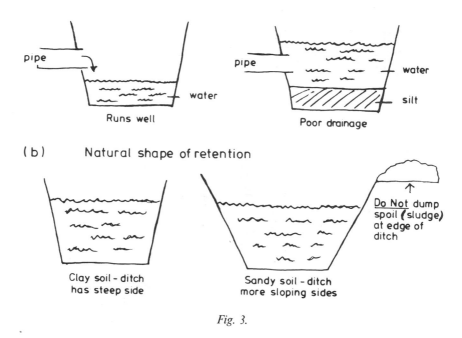

(b)   Natural shape of retention

Fig. 3.

Figure 4 shows a comprehensive drainage calendar.

Drainage systems

There are two types of drainage system:

(a) *Surface drainage,* e.g. rig or ridge-and-furrow, still found on many old pastures in the UK. Levelling this type of field without putting in an alternative drainage system is rarely successful as the system would have been put in for a good reason in the first place! Weeds tend to grow in the furrows, and although the undulating surface may provide good exercise for non-working horses, it can be dangerous if they rush about and also dangerous for heavily pregnant broodmares who may lay down in a dip and get stuck. Another form of surface drainage may be practised in hill areas whereby shallow ditches are cut around the slopes some 20 m (20 yards) apart.

(b) *Under drainage* can be a relatively expensive process, involving the movement of soil, the laying of expensive pipes and filling in afterwards. Typical cost in the UK in 1986 is £620–£645 per ha (£250–£260 per acre).

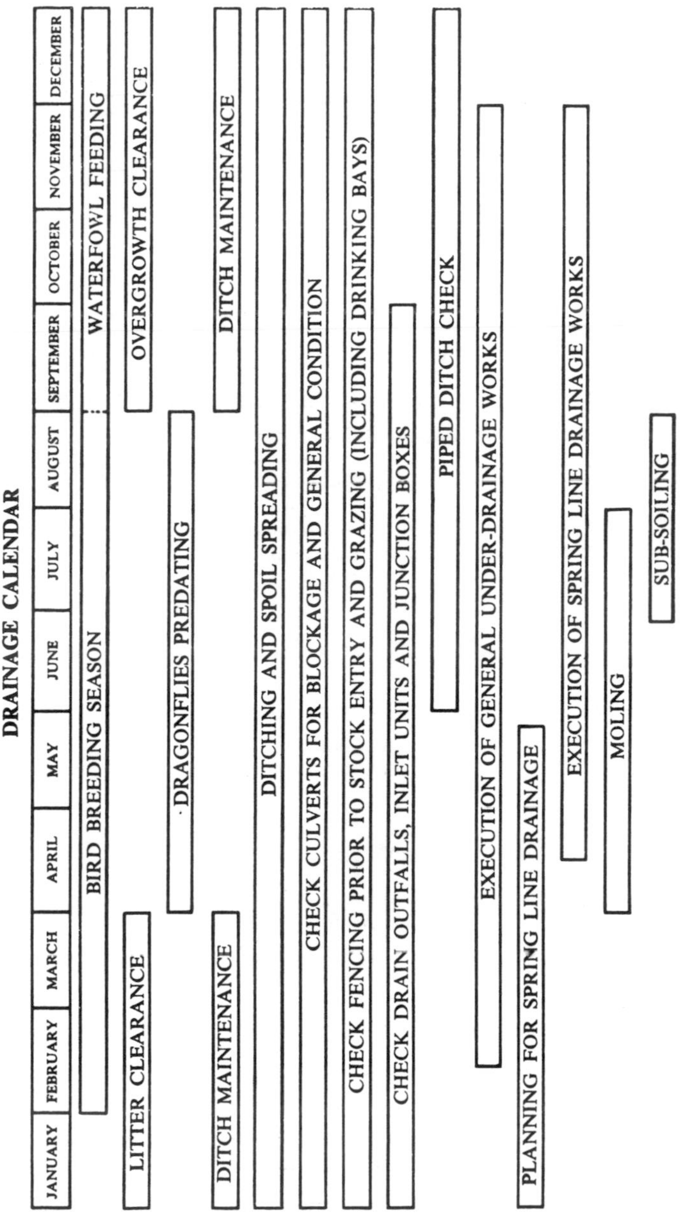

DRAINAGE CALENDAR

	JANUARY	FEBRUARY	MARCH	APRIL	MAY	JUNE	JULY	AUGUST	SEPTEMBER	OCTOBER	NOVEMBER	DECEMBER

WATERFOWL FEEDING

OVERGROWTH CLEARANCE

BIRD BREEDING SEASON

LITTER CLEARANCE

DRAGONFLIES PREDATING

DITCH MAINTENANCE

DITCH MAINTENANCE

DITCHING AND SPOIL SPREADING

CHECK CULVERTS FOR BLOCKAGE AND GENERAL CONDITION

CHECK FENCING PRIOR TO STOCK ENTRY AND GRAZING (INCLUDING DRINKING BAYS)

CHECK DRAIN OUTFALLS, INLET UNITS AND JUNCTION BOXES

PIPED DITCH CHECK

EXECUTION OF GENERAL UNDER-DRAINAGE WORKS

PLANNING FOR SPRING LINE DRAINAGE

EXECUTION OF SPRING LINE DRAINAGE WORKS

MOLING

SUB-SOILING

Fig. 4 Drainage calendar.

However, if you are going to use a tract of land for a number of years the investment should more than repay itself. Ask your State Extension Service (USA) or other government agency (ADAS in the UK) or a private drainage consultant to advise you. There are two main systems of under drainage: natural and parallel (fig. 5).

(a) Natural - uses the slope of the land to give the gradient

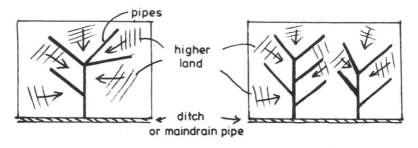

(b) Parallel - used when there is little or no natural slope

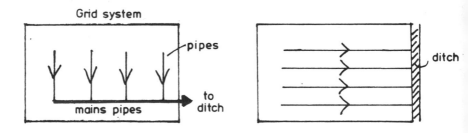

Fig. 5.

The efficiency of the drains will depend on the fall, size, spacing and depth of pipes. Usually a 100 mm (4 in) diameter tile drain will have a slope or fall of 1 in 400. Conventional practice is to install drains at an average of 80 to 100 cm depth. Shallower drains than this can have little effect and deeper drains, except in special circumstances, can be expensive with little extra benefit. It has been shown that poaching of grassland can be significantly reduced if the water table is kept below 50 cm in depth.

Drain spacing affects the rate of water removal. Putting drains very close together does not affect the final result, just the timescale in attaining it. Pipe size should be sufficient to carry excess water efficiently and the pipes must not run under pressure. With traditional 75 mm (3 in) clay pipes, this is rarely a problem but if smaller plastic pipes are used, their capacity for peak flows should be checked.

Permeable fill, usually gravel or similar, is commonly placed in the trench above the pipe to form a permanent connector for moling or subsoiling treatments. It is used mainly with heavy, slowly permeable soils and, although expensive, it ensures that the pipes remain effective for many years, enabling secondary treatment to be repeated when necessary.

Since many soils, especially clayey ones, have low subsoil permeabilities, effective water movement is confined to near the surface. In such soils pipe drains are largely ineffective unless the subsoil properties are modified by subsoiling or moling to form unlined channels to convey water through the soil, via permeable fill, to the pipe drainage system.

The spacing and frequency of tile drains and, therefore, cost may be reduced in heavier soils by use of 'mole' drainage, whereby a 'bomb'-shaped mole is dragged through the soil at right angles to the tiles (or in the absence of tiles) opening directly into the ditch (fig. 6). These mole drains often stay open for three to fifteen years but should be renewed periodically. In 45% clay soils they can last for fifteen years. However, durability of mole drains is largely dependent on climatic conditions, e.g. a drought will break down the mole channels. This technique means the channel is created and not inserted. Gravel free clay soils (at least 35% clay) are needed, with a reasonable slope. The mole channels are made at a depth of up to 750 mm (2 ft 6 in) (ADAS recommend 525 mm or 20 to 21 in), some 2 to 3 m (approximately 3 yards) apart.

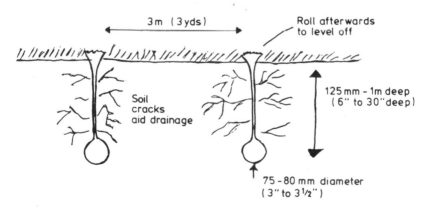

Fig. 6 Mole drains.

Subsoiling aims to break up the subsoil to make it less dense and aid water flow to drains. The normal equipment is similar to a mole plough but it is equipped with a tapered share, commonly with 'wings', to heave the subsoil. Ideally subsoiling should be done when the soil is dry and brittle. The operation needs to be repeated frequently as the soil settles gradually. Many soils are subsoiled when too moist, forming a channel similar to a square mole channel. In a drainage context, in contrast with general soil loosening, the subsoiler is run over pipes which are placed as close as 10 m (10 yards) and rarely more than 30 m (30 yards) apart. The subsoil tines are run through the ground at 1.5 m (1.5 yards) centres and 40 to 50 cm (20 in) depth.

Both subsoiling and moling cause some surface heave, most with subsoiling, and the ground needs to be levelled when under grass, both to allow use of mechanical harvesters and prevent injury to stock. The heave from subsoiling, when done in the correct, dry conditions can be very large.

Subsoiling is very commonly used to improve aeration and loosen topsoils and subsoils in arable farming when the soil does not otherwise require drainage. It should be considered when renovating badly poached grassland. Some apparent drainage problems are caused by poor surface conditions alone and expensive pipe drainage may not be needed. When considering drainage, make sure that expert advice is acquired in relation not only to drainage design, but also to the need for drainage as opposed to simple management.

Note: metric to imperial conversions in this section are not exact, they relate to the pipe sizes and distances used in practice under each system in the UK.

Irrigation

As has been mentioned before, even in the UK irrigation of horse paddocks may be justified, although only if basic husbandry is already sound, and in some parts of the world it is regarded as essential.

Whether irrigation will pay on horse pastures will depend very much on the cost of equipment and its installation, and the cost of the water itself. It may not be regarded as economic to use mains water, but a visit by a competent water diviner (see below) and the discovery of a suitable site for a bore hole on your property may make irrigation a feasible proposition. Check first with your local water authority as the bore hole may alter the water table. Claims made for irrigation of grassland include 'a 73% average increase in dry matter yield over a four year period' and 100–300 kg dry matter increase per acre per 25 mm of water applied, depending on fertilizer usage. If you regularly have a 'grass gap' in high summer due to drought, it may be possible to reduce supplementary feeding or increase stocking rate by judicious use of irrigation.

The problems encountered with irrigation of horse paddocks may be summarised as follows:

• Uniformity of distribution.
• Labour requirement.
• Safety of horses from equipment *in situ*.
• Safety of equipment from horses.

The first aspect obviously relates to the type of system selected. An important point about irrigation equipment is that although it must be bought in *pieces*, it should not be bought *piecemeal*! The irrigation system should be planned and purchased as a unit.

Detailed irrigation management is really beyond the scope of this book, and I suggest you seek the advice of a consultant on the subject, who will take into account your needs for grazing and conservation, labour availability, and the possibility of fixed field boundary systems as a safe method for grazing horses. I suggest moveable systems should *only* be used to get the grass growing, *and then removed before horses are allowed in to graze*.

In the case of both drainage and irrigation, check with an independent consultant before you follow the advice of a contractor whose business is to get you to spend money on his or her system! Two useful (free) booklets from MAFF about irrigation are listed in the reading list at the end of this book, and the equipment is also discussed in Culpin's *Farm Machinery*, which is also listed.

Briefly, the equipment generally consists of a fairly powerful engine (a tractor or, better still, an electric motor in conjunction with a pressure tank and automatic switch gear is ideal), a pump, often of the centrifugal type, delivery pipes and suitable nozzles or jets. The object is to apply the water evenly and not too fast, giving the equivalent of 25 mm (1 in) of rain in three to four hours. Less than half of this at one go is probably a waste of time. Basic systems of distribution include spray lines, small diameter rotary sprinklers and large diameter rotary sprinklers, the latter being most suited to grassland.

Irrigation should not be discounted as an aid to pasture management, especially on studs and intensively managed livery/boarding farms.

Grass and other herbage for grazing

Grass, clover, alfalfa (lucerne) and various 'herbs' are all widely grazed by horses and ponies. Except on specialist legume (clover/alfalfa) pastures, grass or grass/clover mixtures, with or without herbs, they are the most widely used pasture plants for horses.

Grass

Important characteristics are discussed below.

(i) Palatability

Unless they are just there to provide 'bottom' to the sward/turf, pasture grasses are useless if they are not palatable. Very little work has been done on palatability of grasses specifically for horses, and none have been bred (yet) with this specific goal in mind. However, Mrs Mary Tavy Archer has done some preliminary work at the Equine Research Station at Newmarket, and in a paper published in 1980 produced the results shown in table 4.

Table 4 Grasses preferred by horses

Much favoured	Less favoured	Not favoured
Hybrid ryegrass (*Lolium perenne/L. multiflorum*)	**Perennial ryegrass** (*Lolium perenne*)	**Meadow foxtail** (*Alopecurus pratensis*)
Augusta	S23	Timothy
Sabrina	Melle	(*Phleum pratense*)
	Cawdor	S48
	Silian	S50
Tall fescue (*Festuca arundinacea*)	Brown top (*Agrostis tenuis*)	**Perennial ryegrass** (*Lolium perenne*)
Alta		S24
Dovey		
Creeping red fescue (*Festuca rubra*)	**Meadow grasses** (*Poa pratensis*)	**Cocksfoot** (*Dactylis glomerata*)
S59	Kentucky bluegrass	Prairial
Reptans		Cambria
Crested dogstail (*Cynosurus cristatus*)		**Meadow fescue** (*Festuca pratensis*)
		S53

Only those grasses actually tested are included in this table Archer, M. T. (1973, 1978, unpublished); Archer, M. T. (1980) *Veterinary Record*, 23 August

It is interesting that the famous Kentucky bluegrass was designated as 'less favoured', although this may have something to do with UK soil conditions. Some of the specific varieties mentioned are likely to become unavailable as seed merchants, in their quest for high grass yields for dairy cows and beef, (especially with the EEC introduction of milk quotas and the swing back to 'milk from grass' systems), introduce new strains and varieties. It is to be hoped some of this will prove of use for horses. In fact I understand that the grasses were grazed at a fixed and constant height, which could make the results somewhat unrepresentative as different varieties could be at different growth stages, which affects palatibility, at that height.

Mrs Archer found that there were differences between individual horses' preferences for various grasses, but no clearcut differences between breeds of horses. 'Despite careful chemical analysis, no satisfactory explanation for these preferences has been established.' Mrs Archer also pointed out that 'It is not true that horses like only short grass but rather that the preferred grasses never have a chance to grow long where horses are grazing regularly'.

(ii) Persistence

Persistence is the ability to withstand hard grazing and cutting.

(iii) Winter hardiness

(iv) Resistance to disease

Resistance to disease is not a major consideration with current varieties, although some diseased herbage may be implicated in abortion and infertility with broodmares due possibly to the presence of mycotoxins (see below).

(v) Heading date

Heading date is the time of seeding – coming into flower. This is most crucial for grasses for conservation (hay/silage) but is relevant for grazing as it determines the timing and patterns of growth periods. A sward made up of grasses of varying heading dates will ensure continuity of production throughout the growing season. Heading dates for most varieties can be predicted fairly accurately – within three to four days each side. Figure 7 illustrates growth stages in grass.

(a)

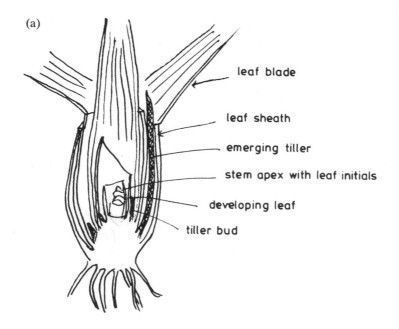

leaf blade

leaf sheath

emerging tiller

stem apex with leaf initials

developing leaf

tiller bud

(b)

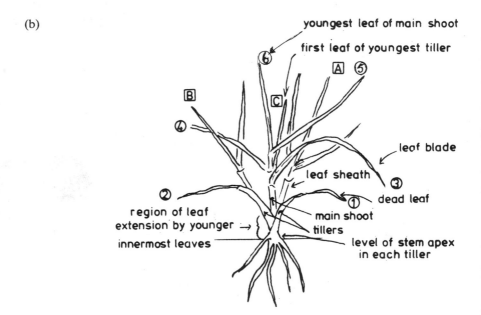

youngest leaf of main shoot

first leaf of youngest tiller

leaf blade

leaf sheath

dead leaf

region of leaf extension by younger →

main shoot

tillers

innermost leaves

level of stem apex in each tiller

Fig. 7 (a) Cross-section of vegetative grass plant showing stem apex buried in the leaf sheaths; (*b*) Vegetative grass plant, showing tiller growth (A–C) and succession of leaves (1–6). (*After O. R. Jewiss GRI.*)

(vi) Yield

Yield is calculated in terms of dry matter. For grasses for conservation, high yields are desirable, but for grazing by horses, lower yields may be more easy to manage, which is partly why the best 'dairy cow' grass varieties are not always ideal for horses.

(vii) Digestibility

Digestibility is the percentage of nutrients present in the grass which the animal can actually extract. The more fibre (cellulose/lignin) in the grass, the less the horse can digest. However, a certain level of fibre is desirable to maintain a healthy digestive tract, and give a feeling of 'repletion and contentment' as will be noted when horses and ponies on lush spring grass start barking trees and chewing fences. Some hay or straw may be offered with advantage at such times. Conversely, too much fibre on old, headed grass may make the pony look 'fat' to the untutored eye; that is he develops a pot or hay belly, but really he is filling himself up with rubbish in an effort to obtain the vital nutrients he is craving. The hind gut becomes distended as the pony's digestive micro-organisms take a long time to ferment the woody fibre. This is most likely to occur in autumn after the 'autumn flush' and many owners make the mistake of thinking a field has plenty of 'keep' when there looks like a lot of grass on it – but in fact it is mostly indigestible rubbish.

Grass species and mixtures

All grasses have a basic growth curve similar to that shown in fig. 8. The ideal grass mixture will be such as to level out the peaks and troughs, by choosing a mixture of early and late growing grasses for herbage early and late in the year with staggered heading dates. Hard feed and conserved forages may then be used to correct any deficiencies, and in fact both may be required in winter, then just forage (for fibre) in spring, through to concentrates in autumn. Feed levels should be sufficient for maintenance, with additions for growth, pregnancy, lactation and work where appropriate. This is a basic pattern and the timing of the peaks will vary considerably in different areas and with different grass varieties. Figure 8 also gives an indication of the effect of rainfall on dry matter yield.

In general terms, in temperate countries a basic horse paddock mixture will contain:
(a) *two* penennial ryegrass (*Lolium perenne*) varieties to make up 50% of the mixture; choose from: prostrate, late-heading varieties (such as

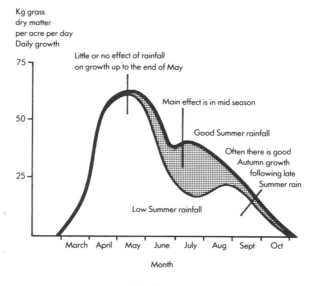

Fig. 8.

S23, Melle, Perma, Meltra, Petra, Semperweide, Endura etc.), plus
(b) *two* creeping red fescues (*Festuca rubra*) to make up 25% of the
    mixture (such as S59, Reptans, etc.);
(c) The remaining 25% being made up of
    (i)   crested dog's tail (*Cynosurus cristatus*) (5–10%)
    (ii) rough stalked meadow grass (RSMG, on wet areas) at (5–10%),
          or smooth stalked meadow Grass (SSMG, on dry areas) at
          (5–10%), plus
    (iii) a little wild white clover (1–2%).

Some Timothy (S48) may also be desirable, especially if hay is to be taken,
and possibly a late-heading variety of cocksfoot (known as 'orchard
grass', *Dactylis glomerata*, in the USA) if drought is a problem, although
it is not always well liked by horses.

The use of mixtures of varieties of different grass species is desirable
both from a palatability viewpoint and for preventing the spread of plant
diseases. It can also extend the usefulness of the pasture by spreading
growth patterns. It may be useful to put a well-drained paddock down to
an 'early' mixture to give grazing in early spring. *Agrostis species* may also
be added into the mixture where the paddock is to be used for both
exercise and grazing, as it stands up well to 'wear and tear'. Agrostis can
also be used in gateways and around water troughs, provided it is not
allowed to invade the main pasture area, and on gallops and race courses.

Examples of some commercial UK grass seed mixtures sold specifically for horse paddocks are as follows:

RHM grass mixture for stud farm paddocks

'This mixture is designed for Stud Farm Paddocks for brood-mares, to produce a closely knit sward, with a long life expectancy. It can be utilised alone or in conjunction with beef cattle to minimise the worm burden.'

10% Certified Highland bent grass
10% Certified chewings fescue
15% Certified ruby creeping red fescue
15% Certified Dawson creeping red fescue
15% Certified Aberystwyth S48 timothy
35% Certified Angela (extreme pasture perennial ryegrass)
100%

Recommended seed rate 40–50 kg/ha (36–45 lb/acre approximately). (Mid 1970s mixture.)

RHM grass mixture for horse and pony paddocks on farms in the UK

'Designed as a permanent paddock, this ley will meet the nutritional grass needs of horses or ponies and will at the same time provide an excellent exercise area. A very small amount of wild white clover is included, but should this not be required, we suggest raising the Highland bent grass to 10%.'

2.5% Certified wild white clover
7.5% Certified Highland bent grass
10 % Certified Aberystwyth S48 timothy
30 % Certified creeping red fescue
50 % Certified Angela (extreme pasture perennial ryegrass)
100 %

Recommended seed rate 33–37 kg/ha (30–33 lb/acre approximately). (Mid 1970s mixture.)

Miln Marsters thoroughbred mixture (1986) UK

'This mixture will suit many thoroughbred studs in the British Isles and once established, may be regarded as permanent pasture.'

    10 g Kent wild white clover
    55 g Crested dogstail
    55 g Rough stalked meadow grass
    55 g Elite herb blend with dandelion (uncertified)
  110 g Condesa (tetraploid) late perennial ryegrass
  165 g Manhattan turf perennial ryegrass
  165 g Trani late perennial ryegrass
  165 g Boreal creeping red fescue
  <u>220 g</u> Parcour late perennial ryegrass
    1 kg

Suggested sowing rate approximately 18.25 kg/acre.

Miln Marsters horse and pony mixture (1986) UK

'A more general purpose mixture to provide grazing and an occasional cut of hay.'

  17.5 g Kent wild white clover
  40   g Crested dogstail
  40   g Smooth stalked meadow grass
  72.5 g Elite herb blend with dandelion (uncertified)
  150   g Manhattan turf perennial ryegrass
  150   g Magela medium perennial ryegrass
  150   g Melle late perennial ryegrass
  150   g Boreal creeping red fescue
  <u>230   g</u> Meltra late perennial ryegrass (tet)
    1 kg

Suggested sowing rate approximately 13.25 kg per acre.

Miln Marsters renovation mixture

'As an alternative to ploughing and reseeding, it may be possible on some soils to oversow with a turf forming grasses mixture.'

  125   g Smooth stalked meadow grass
  250   g Boreal creeping red fescue
  312.5 g Manhattan turf perennial ryegrass
  <u>312.5 g</u> Melle late perennial ryegrass
    1 kg

Suggested sowing rate approximately 8 kg per acre.

Hunters of Chester cardi mixture – 98 (long-term intensive)

Hunters of Chester are well-known for their horse paddock mixtures, including their 'paddock special' 1 acre packs which are even sold in some tack and horse sundries shops.

```
0.5 lb Certified grasslands Huia white clover
1   lb Certified Kersey white clover
1.5 lb Certified Aberystwyth S48 timothy
2   lb Certified Farol timothy
4   lb Certified Augusta hybrid ryegrass
4   lb Certified Meltra tetraploid perennial ryegrass
4   lb Certified creeping red fescue
6   lb Certified Melle perennial ryegrass
9   lb Certified Aberystwyth S23/Wendy perennial ryegrass
```
33 lb/acre (1 lb = 0.45 kg)

Finney Lock HM1 UK 1981

'This is a long term general purpose mixture which will be suitable for cutting and grazing and will stand up to extreme conditions. It is both persistent and winter hardy and the varieties in it spread grazing over the longest period possible. During its development it should be grazed carefully but often. In its second year this mixture can give excellent hay cuts providing a quality product which is ideal for horses.'

```
0.5  kg Timothy Aberystwyth S48
0.5  kg Clover white N.Z. grasslands Huia
0.5  kg Clover white Sonja
1    kg Ryegrass perennial Aberystwyth S321
1    kg Cocksfoot Cambria
1    kg Clover red late-flowering Aberystwyth S123
1.25 kg Ryegrass perennial Aberystwyth S23
1.5  kg Ryegrass Italian Aberystwyth S22
1.5  kg Cocksfoot Aberystwyth S26
1.5  kg Timothy Kampe 11
2.5  kg Ryegrass perennial Aberystwyth S24
```
12.75 kg per acre

Finney Lock HM2 UK 1981

'This is a long term grazing mixture especially suitable for paddock grazing. Stands up to terrific punishment and comes back for more.'

    0.5 kg Clover white N.Z. grasslands Huia
    0.5 kg Clover white Sonja
    1.5 kg Timothy Aberystwyth S48
    2.5 kg Ryegrass perennial tetraploid Taptoe
    2.5 kg Ryegrass perennial Talbot
    <u> 4   kg Ryegrass perennial Melle</u>
   11.5 kg per acre

'The different species used to formulate these mixtures have been specially chosen to ensure that both leys are palatable through the whole season.'

Finney Lock hay and grazing

    0.5 kg Wild white clover Aberystwyth S184
    1.5 kg Hybrid ryegrass grasslands Manawa
    2   kg Creeping red fescue Reptans
    6   kg Meadow fescue Aberystwyth S215
    <u> 6   kg Timothy Aberystwyth S48</u>
   16   kg per acre

Finney Lock grazing

    0.5 kg Wild white clover Aberystwyth S184
    1.5 kg White clover NZ Huia
    <u> 8   kg Perennial ryegrass Aberystwyth S23</u>
   10   kg per acre

These mixtures are all given as examples of different seed houses' interpretations of the requirements for horses. Appendix 1 gives some mixtures suggested for gallops, race courses and exercise paddocks.

The basic pasture characteristics of the main chosen grass species (fig. 9) are as follows:

(a) Perennial ryegrass (*Lolium perenne*) also known as English ryegrass, Pacey's Devonshire evergreen (!), Leafy Lurgan, Raygrass and

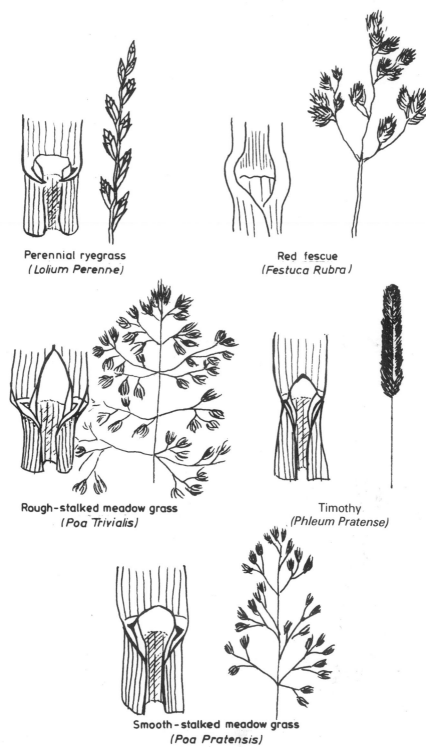

Perennial ryegrass
(Lolium Perenne)

Red fescue
(Festuca Rubra)

Rough-stalked meadow grass
(Poa Trivialis)

Timothy
(Phleum Pratense)

Smooth-stalked meadow grass
(Poa Pratensis)

Fig. 9 Desirable pasture grasses.

Eavers. Perennial ryegrass is very persistent in good rich soil, but tends to decline after two or three years on poor light soil unless it is kept well fertilized. It grows nearly everywhere in the British Isles and is now found worldwide, forming the basis of many herbage seed mixtures. It is very versatile and does well under all types of conditions.

(b) Red Fescues (*Festuca rubra*). These vary considerably in quality but the better varieties are palatable and of good nutritional quality. They are especially valuable under difficult conditions and on hill land. The seed is somewhat expensive because of demand for sport turf, and as establishment is poor it is needed at relatively high seed rates. *Festuca rubra genuina* denotes the creeping type.

(c) Rough stalked meadow grass (*Poa trivialis*). This highly palatable species is best used on moist, rich soils in sheltered conditions. It gives good 'bottom' to a sward having a close growing habit and is highly nutritive. By filling up the bottom of a pasture it also helps to keep out undesirable species.

(d) Smooth stalked meadow grass (*Poa pratensis*), also known as Kentucky bluegrass. This grass is able to withstand severe drought thanks to its creeping underground stems, so it is especially useful on light, dry, sandy soils.

(e) Timothy (*Phleum pratense*), also known as catstail, was first cultivated in the USA by Timothy Hansen and introduced to the UK in 1720. Timothy produces abundant grass seed which means it is cheap, but excessive amounts should not be used in grazing mixtures. However, pure timothy hay is much favoured for horses. It prefers a moist soil – heavy loams and clays, and peaty soils.

(f) Wild white clover (*Trifolium repens*) (not illustrated). This perennial, also known as 'Dutch' clover, grows on almost every kind of soil. It is a deep rooted plant and withstands the effect of drought on dry, sandy soils. Micro-organisms in nodules on the root system 'fix' nitrogen from the air (see Appendix/Glossary) which enables it to enhance the overall soil fertility. Most farm varieties are broad-leaved, aggressive forms and will 'take over' on paddock so true wild varieties should be selected.

However, clover can present a problem by supplying horses with excess nitrogen, and therefore protein, in pasture, and although a small amount of wild white clover (say 1–2% of the seed mixture, by weight) is desirable, especially for breeding stock, excessive amounts should be avoided. (Pure clover, i.e. red clover, swards are grazed by broodmares in some parts of the world, but a very high standard of management is required if this practice is to be successful.)

Wild white clover does *not* like heavy doses of nitrogen fertilizer, and indeed this treatment can be the most effective means of reducing the clover population in a sward (followed by very hard grazing by cattle before putting horses back on, as the grass will be lush from the nitrogen). Several treatments may be required over more than one year to rebalance the plant population of the sward in favour of horses. Horses and ponies do not appear to find red clover (*Trifolium pratense perenne*) palatable given the varieties which are currently available, although I understand it is widely grazed on breeding farms in the USA.

Some horses suffer from a form of photosensitisation in a set of circumstances involving clovers or other trifoliates (alfalfa and possibly fenugreek), and the condition is frequently misdiagnosed as mud fever. The combination of circumstances involved is (usually) a chestnut horse (sorrel, through bright- to liver-chestnut) with white on legs/face, eating trifoliates (in grazing, hay, silage, hay-age, moist vacuum-packed forage or dehydrated, such as lucerne or alfalfa nuts, meal or wafers, incorporated in concentrates or herb mixes, etc.) and being wormed with thiabendazole wormers.

A typical example, belonging to a client of mine, is a valuable thoroughbred brood mare which is a dark chestnut with two white socks and a white blaze. She suffered extensive hair loss and scabs exuding pus, on the white fetlocks and face only, and she was diagnosed by the vet as having mud fever, and had been dressed with cortisone cream for weeks with no effect. She was being fed alfalfa/timothy (i.e. trifoliate containing) hay at the time, and had been wormed with a thiabendazole anthelmintic. Within days of stopping the alfalfa/timothy hay (exchanging it for a clover-free Yorkshire meadow hay) the condition had cleared up. When she was accidentally given the wrong hay, once, some weeks later, her leg immediately 'blew up' and took ten to fourteen days to go back to normal size.

This is *not* a common problem but is, I believe, often wrongly diagnosed and treated in susceptible horses, and horse keepers and vets should be aware of the possiblility of it occurring. It is possible that a predisposition to this condition is hereditary. I have not, as yet, come across its occurrence in these circumstances in non-chestnut horses, although it may occur in strawberry roans or chestnut skewbalds (pinto, 'paint' or 'coloured' horses in the USA).

Certain herbage varieties, whilst not being strictly poisonous, can inhibit production in some way. For example, the fescues are often used to give 'bottom' to the sward, along with drought resistance. In the USA, it is estimated that fescues, first cultivated in the 1930s, are now to be found growing over an estimated 14 million ha (35 m acres).

Unfortunately, fescue pastures and hay, particularly when the sward is pure fescue, can lead to reproductive difficulties in horses. The most common symptoms of fescue toxicity are little or no milk production (agalactia) in newly foaled mares, exceptionally thick placentas which have been known to suffocate new-born foals, and gestation periods which may be prolonged by more than two weeks. There is also thought to be a possible association between fescue toxicity and abortions and stillbirths.

The grass itself does not in fact cause the problems, but it is host to a parasitic fungus called *Epichloe typina*, which causes a toxic reaction in animals eating it. This fungal infection is apparently quite widespread in fescue swards throughout the USA, and will be discussed further in the section on *mycotoxins* (p. 103).

Another problem associated with herbage varieties is that of *photosensitivity* after feeding *alsike* clover (a trifoliate), which is occasionally fed in the UK, and widely used in the USA. This seems to occur whether or not thiabendazole wormers are used. Grey horses or those with white on them apparently tend to be affected, and may break out in running sores in the white areas, again misdiagnosed as mud fever or rain scald. Some horses also exhibit noticeable nervous reactions. The hooves may also be affected with cracks and separation between the hoof and the coronary band.

The problem usually arises after prolonged grazing on alsike clover. The alsike contains a toxin which damages the liver so that it cannot drain properly into the intestinal tract, and this obstruction causes the metabolic by-products of chlorophyll (which is the green pigment contained in all green plants) to 'spill' back into the blood. When white parts of the horse are exposed to direct sunlight, these 'photodynamic' chlorophyll byproducts react by causing skin cells in the area to die. If the obstruction becomes severe enough it can be fatal. Alsike is thought to be most highly toxic in the flowering stage of growth. Excess selenium (Se) or copper (Cu) in the soil can cause liver damage which can make horses more susceptible to the dangers of alsike. However, alsike tends to accumulate the trace element molybdenum (Mb) which can help to counteract the effects of excess copper, so provided molybdenum levels in the soil are adequate, selenium levels, especially in alkaline soils, are the most crucial ones.

The toxic principle in alsike, has not, as yet, been identified, but it seems likely that horses in poor condition, especially those pastured on soils with trace element imbalances, are likely to be the most at risk. Cattle are less susceptible than horses to alsike poisoning.

Another potential problem which *may* be associated with grazed herbage is the oestrogenic activity of lucerne (alfalfa) and certain clovers.

I can find no specific evidence of problems in this context, and both species are widely grazed, and fed as hay to breeding stock. However, I do wonder whether this *may* be associated with abnormal oestrus in breeding cycles. In fact, a closely related plant, fenugreek, is grown specifically to extract constituents for the human contraceptive 'pill' although paradoxically fenugreek does *not* show oestrogenic activity!

Herbs in horse pastures

The value of introducing herbs to horse pastures is somewhat debatable. Nutritionally herbs can make a valuable (if relatively unmeasurable) contribution to the horse's diet, particularly its mineral status, and this may be especially important for ponies or on pony studs, where concentrate feeds are restricted in use.

Herbs tend to exhibit 'luxury uptake' of certain minerals, making them a rich source of these nutrients, and they are also highly palatable. Horses led out and grazed along verges are often quite insistent about eating specific plants, which may or may not be indicative of a craving for the minerals they contain, or just of a liking for the taste. It is true that many horses fed a mineral supplement for the first time lose interest in eating herbs (and soil, droppings, tree bark, etc.).

Fertilizer treatment of pastures, even using organic fertilizers, tends to disfavour the herbs, which often prefer less fertile growing conditions. On intensively grazed, managed pastures I am, therefore, inclined to leave herbs out of a seed mixture, and ensure the rest of the diet is balanced for minerals. On the whole, I prefer to feed minerals in the hard feed, or feed blocks/licks, and be reasonably sure the animal is getting something like what it needs, than to rely on herbs, and a particular horse having a taste for the 'right' herbs. If for some reason grass kept animals are not likely to receive supplementary feeding, as a long-term policy, then I should include herbs in the herbage seed mixture, or sow herb strips perhaps on the headlands if these are not excessively trampled (plate 5). Conservationists will of course favour maintaining a herb population in the pasture.

Acceptable pasture herbs

Figure 10 illustrates acceptable pasture herbs, some of which are described in more detail below.

Dandelion (*Taraxacum officionale*)

Also known as 'fairy clock' 'wet-the-beds' (*pisse-en-lit*, France), dandel-

Plate 5 This herb strip contains yarrow (centre), plantain (left), dandelions and clover. The presence of various weeds illustrates the problem of managing such areas, as most herbicides would kill the herbs as well.

ion is a member of the Compositae, found in Europe, North and Central Asia, North Africa and North America. It is poor in protein and glucosides, with virtually no fats, but rich in potassium and calcium salts, manganese, sodium, sulphur, vitamins A, B, C and D and choline. It is a diuretic and mild laxative (hence its 'common' names). It produces ethylene gas which can inhibit growth of other plants, so it may not be desirable in a seed mixture containing plants which are difficult to establish. (Interestingly, ethylene speeds the ripening of fruit and some orchard owners are now planting dandelions under their fruit trees!) The name dandelion comes from the French dent-de-lion, referring to the backward slanting toothed shape of the leaves. The bright yellow flower (which closes at night) and fluffy seed-head (used as 'clocks' by countless school children) need little description.

Narrow-leaved plantain (*Plantago lanceolata*)

Narrow-leaved plantain is also known as English plantain, waybroad, waybread and ratstail, and in the antipodes 'Englishman's Foot' (planta means sole of the foot) because it was said that everywhere an Englishman trod, a plantain sprang up!

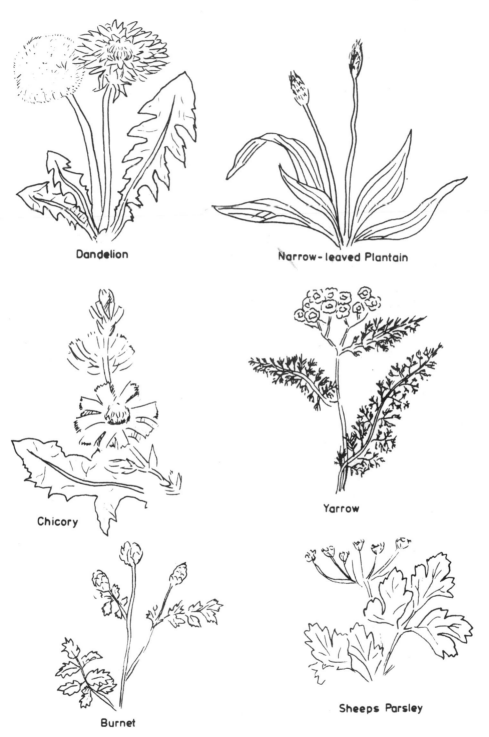

Fig. 10 Acceptable pasture herbs.

This perennial is rich in minerals, especially potassium, calcium and sulphur, and contains some vitamin K. Personally I would regard this as a weed except on poor, light, elevated pastures where it can produce some useful herbage. It has an abundant growth habit, but tends to encroach on surrounding grasses and 'take over'. It is a very persistent plant but, although it does not look particularly succulent, its leaves contain so much water that when made into hay it loses more weight than any other foliage plant! Often included in pasture herbage mixtures, but I would suggest a very low level.

Chicory (*Cichorium intybus*)

Also called succory, blue sailors and our lady's eyes. Do not confuse this useful plant with the vegetable endive, which for some reason is also known as chicory.

Like the dandelion, *Cichorium intybus* is a member of the Compositae. It is a native of Eurasia and North Africa, but is naturalised all over Western Europe. The ground roots are used in coffee, not as some say to 'adulterate it' but to protect the liver (because it contains choline) against the harmful affects of the caffeine in the coffee. However, here we are interested in the leaves. The variety known as Belgian or Brussels chicory is the most extensively grown in Europe, even grown on its own as a rich fodder crop for cattle in some parts of France and Northern Italy.

Chicory is a rapid growing biennial, with bright blue cornflower or daisy-shaped blooms, and grows best on light soils on gravel and chalk. It is very deep rooting, and the leaves are rich in inulin, fructose and choline. It does not like fertilizer so there is no point in sowing it where you plan to use fertilizer extensively.

Yarrow (*Achillea millefolium*)

Also called wound-wort, soldiers' herb, staunch grass, sanguinary, millfoil and thousand leaf. Yarrow (from the Anglo-Saxon *gaerwe*) derives its generic name from Achilles who, following the instructions of the wise centaur Cheiron, staunched the wounds of his soldiers with this herb.

Yarrow is another member of the genus Compositae and thrives on the poorest and driest soils, as well as on those which are heavy and wet. Most horses love the taste, and the plant is said to enhance the flavour, 'strength' and health of other plants grown near it, presumably because it is so deep rooting. This latter quality also means it is very drought resistant.

Its medicinal properties have been recognised since time began and if I were going to include just one herb in the pasture, this would be the one I would choose. Unfortunately, the seed is very expensive, but as there are some eight million seeds to the kilogram and a reasonable inclusion rate would be about 250 g/ha (100 g/acre), this is not too disadvantageous.

Yarrow is strongly scented, with tiny daisy-like flowers growing in high clusters.

Burnet (*Poterium sanguisorba*)

Often recommended for sheep grazing, this deep-rooting, drought (and flood) resisting perennial plant thrives on any land, but most particularly dry soils. It is a palatable herb which stays green all year round and is supposed to be a general 'blood purifier' and 'tonic', whatever *that* may mean! It seems to grow better in strips than as single isolated plants.

Sheep's parsley (*Petroselinum crispum*)

Very palatable but not very persistent under grazing conditions and I personally would not bother to include it in a horse paddock herbage mixture.

Wild garlic (*Allium oleraceum* – field garlic and *Allium ursinum* – wood garlic)

There are various varieties of wild garlic found in verges, hedgerows and in the fields, and these are relished by horses. Eating them is thought to repel flies and protect against colds and viruses.

Plate 6 Most horses love wild garlic and consuming it seems to help keep the flies away. Only plant it where you want it as it spreads like wildfire.

Comfrey (*Symphytium officinale*)

Comfrey is a useful herb for feeding to horses but is not normally grazed. (See section on comfrey, starting p. 51.)

Other grazed crops

Lucerne, fenugreek and clover may also be grown for grazing, but are more likely to be used for conservation (see next section).

Grass and other herbage for cutting and conservation and hand feeding

The conservation crop

Different types of herbage may be grown specifically for harvesting and feeding as fresh or conserved forages to horses. In the UK, in view of the changes in EEC farming policy, many farmers are seeking more profitable use of their grassland and are turning to growing hay, hayage, silage, MVP (moist vacuum-packed) forage and dehydrated grass or alfalfa specifically for the horse market. Racehorse trainers and studs are prepared to pay high prices for top quality forage and, mystifyingly, will also pay grossly inflated prices for imported forages which often prove to be of very inferior nutritional quality indeed.

It seems likely that in the next few years a number of farmers will turn towards growing 'horse-hay' and other forages as specialist producers, and they should find a ready market. It is beyond the scope of this book to talk in detail about forage conservation but a number of factors may be worthy of consideration:

(a) Some horse owners and stud managers are, rightly or wrongly, concerned about the high level of nitrates occuring in heavily fertilized forage crops, and a specialist grower prepared to put some effort into marketing his crop may find it worthwhile to use organic or semi-organic fertilizers for horse hay.

(b) Alfalfa, clover and sainfoin hays may be preferred for the same reason because, although a little nitrate fertilizer may be applied in the year of establishment of the crop, these legumes 'fix' their own nitrogen from the air with the aid of micro-organisms in their root nodules. Very little nitrate or nitrite is likely to be present in the harvested crop. However, because the crop fixes its own nitrogen, and nitrogen is an essential component of protein, legumes (clover, sainfoin, alfalfa) tend to be richer in protein than grasses, and in general contain more protein than mature horses, other than broodmares, in fact require. They do, however, make valuable feeds for young animals, particularly those destined for high performance, such as racing as two year olds.

In general, alfalfa grows best only in the south-eastern parts of Britain, whereas it is grown more extensively in the USA and Australia. Sainfoin waxes and wanes in popularity, and perhaps an upsurge may be expected in the search for 'new' crops.

(c) Another legume which may well 'catch on' is fenugreek, which is similar to alfalfa, but varieties have been developed which are more tolerant of the varying British growing conditions. Fenugreek, or 'Greek hay', has been widely fed to equines since Roman times, and is still commonly fed by farmers surrounding the Mediterranean, and in Malta and Greece. It is popular because it is highly palatable and is often used to disguise forage of inferior quality. This is *not* a good idea, however.

Fenugreek is grown for its seeds which are used as a spice (e.g. in curry powder and animal feed flavouring) and as a source of raw materials for producing the contraceptive pill. It may well gain popularity as a 'cash crop' leading to increased availability of threshed fenugreek hay, but I also believe it is worth looking into specifically as a high class forage crop for horses. In the UK varieties were developed by Dr Roland Hardman of the Department of Pharmacy and Pharmacology at Bath University, and the National Seed Development Organisation (NSDO), whose address appears at the end of this book.

(d) Specialist growers of forage for the horse market may feel justified in using semi-organic or 'biochemical' fertilizers, both to restrict the use of inorganic nitrogen (nitrates), and because these fertilizers also supply a certain level of trace elements which may be of benefit both to the crop itself and to the animals eating it.

(e) Facilities for barn hay drying would be prudent for the grower wishing to produce consistently top quality hay for the horse market.

(f) Preservatives should be given due consideration, particularly when conserving big-bale bagged silage/hayage for horses.

Considerable adverse publicity arose in the UK in 1984 when a number of horses died of botulism after being fed big-bale bagged silage, specifically prepared and marketed for horses without the use of preservatives, which was offered as a sales point, the manufacturers having gathered that horse keepers can be rather touchy about preservatives. In fact, had a preservative been used, those horses would probably have been alive and well today.

It seems that the organism causing botulism, *Botulinum chlostridium*, is endemic in soils, but only produces the endotoxin which results in disease symptoms under alkaline conditions. It would appear that the disease does not arise if the forage has a pH of less than 5.5 (pH 0–7 is decreasingly acid, 7 is neutral and 7–14 is increasingly alkaline). This is why, when home canning of fruits and vegetables was popular, canned fruits which are acidic were generally safe, but home-canned vegetables which are less acidic sometimes caused human fatalities due to botulism.

So, if you plan to feed big-bale bagged silage, or indeed any silage, ensure the pH is below 5.5, applying a preservative if necessary to achieve this aim.

One of the reasons horse owners have not been keen to feed forage treated with preservatives has been the lack of information about their use in horse feeds. However, studies at the University of Illinois and Cornell University indicate that proprionic acid treated hay can be fed to horses, although a two to three day adjustment period may be required if the hay is to be readily accepted, which is reasonable for any new feed source. Proprionic acid is a volatile fatty acid (VFA) which occurs naturally in the horse's hind gut as a result of microbial fermentation of roughages in the diet, so it is not alien to the horse. Other hay preservatives are now available which are considered to be more effective, but there has been no trial work to establish their safety for horses.

Hay treated with proprionic acid (80%) and acetic acid (another VFA) (20%) could be baled at up to 27% moisture content (untreated hay must have less than 20% moisture content if spoilage is to be avoided), so better hay can be made in poor harvesting (weather) conditions. Results quoted by Harold Hintz at the 1985 British Equine Veterinary Association Congress, indicating improvements in feed production efficiency, are shown in table 5. Although a potentially poorer (wetter) crop, the treated hay proved to be more efficiently saved and of a better feeding value than the conventionally harvested crop. The researchers found that, given a choice, horses would eat initially only half as much treated alfalfa hay as untreated, but given no choice they ate the same amount of treated hay quite happily within two to three days. As with proprionic acid treated cereals, it is important to ensure there is adequate vitamin E in the diet.

An alternative preservative available in the UK, which is likely to be very palatable to horses, is a mixture of cider apple vinegar (CAV), which is acetic acid, and molasses, although I have no experience of its use and effectiveness as a preservative.

Taking a hay crop from grazing land

These procedures may vary from taking a hay crop as an afterthought from grazing paddocks which have 'grown away' and are surplus to requirements, through to a planned programme of shutting up paddocks for conservation after spring grazing. A decision, based on the growth rates and grazing requirements, will need to be made, and once the field is 'shut up' an application of fertilizer is indicated to encourage a heavier crop of higher feed value, at least a month before the expected harvest date.

If paddocks are to be regularly used for grazing combined with conservation, it makes sense when reseeding or upgrading the pasture, to include some favoured grass varieties for conservation including for example, timothy grass. It is also vital to ensure that any poisonous weeds, especially ragwort (which is not normally eaten whilst growing, but becomes highly palatable when wilted or dried as it is in hay), are eradicated and removed before the crop is harvested.

Table 5 Effect of proprionic acid treatment of hay on feed production efficiency

	Moisture content at harvest (%)	Field losses (%)	Feed losses (%)	Total losses (%)
Acid treatment	27	10	6	16
Untreated hay	17	17	13	30*

*Not including rain damage

Comfrey

The other non-arable crop which the ordinary horse owner might get involved in cultivating is comfrey, a valuable herb which can be used as a succulent forage plant. It can be chopped into the concentrate feed or fed in haynets, when it is usually better accepted by allowing it to wilt for a couple of hours after cutting (plate 7).

Full discussion of the feeding attributes of comfrey are beyond the scope of this book, but in general terms it is a protein and mineral rich plant which appears to enhance wound healing and, possibly because of its allantoin content, is said to ease arthritic conditions. Quite apart from this it is a useful palatable succulent for the stabled horse. It appears to have similar feeding benefits to yucca, which is increasingly fed in

Plate 7 Comfrey can be an extremely useful supplementary feed for both pastured and stabled horses. It is best fed slightly wilted. (*Photograph supplied by the Henry Doubleday Research Association.*)

powdered form in the USA, and has recently been introduced to the UK as a constituent of vitamin/mineral supplements. Although the two plants are very different, as herbal 'treatments' they seem to be indicated in the same sort of situations.

A number of thoroughbred studs in the UK grow comfrey in the gaps between each paddock, and cut a supply daily and drop it over the fences each side for the mares and youngstock. Comfrey is often found on verges, and plants can be grown in a rough corner of the garden, behind the stables, etc. However, you should only plant it where you want it to grow as once it is established, it is very hard to get rid of! For larger stables a proper comfrey patch may be established, and leaves cut daily from the plants for feeding to stabled or pastured horses.

It is a good idea not to overestimate the requirements for comfrey. Between 1880 and 1890, yields of between 80 and 100 tons of the Russian variety per acre were not uncommon. I would suggest that no more than two plants per three horses would be required.

The comfrey strain most commonly grown for horses was the Stephenson strain developed in 1938 and named for E. V. Stephenson of Hunsley House Stud, Little Weighton, near Hull, England, who had the oldest cared for plot in Britain, from which he harvested satisfactory yields for thirty-six years, although the land was considered poor. On the whole, horses seem to prefer the 'Bocking no. 14' variety of the Stephenson strain or the 'Bocking mixture'. Yields of 30–33 tons per acre should be easily obtainable.

Comfrey is in fact the common name given to the genus *Symphytum*, which is a member of the *Boraginaceae* (borage) family. A native of Eastern Europe and Western Asia, this is a 'coarse perennial herb up to 2.6 m (five feet) high, leafy, branching, with hairy stems and leaves; the lower leaves are generally larger than the upper ones; extensively fleshy root system; flowers – whitish, yellowish or dull purple; generally adapted to moist, cool places.' Cultivated comfrey is often referred to as Russian or Quaker (in the USA and Canada) comfrey. It is propagated by root cuttings, or root 'off sets' and crown cuttings or crown 'offsets' because seed takes about four years to make a plant as good as a one year old from a root cutting.

Growing comfrey

Comfrey should always be planted on clean land from which weeds, especially persistent grass weeds such as couch grass (*Agropyron repens*, also called wickens, squitch, etc., in Britain, or kikuyu grass in Africa, etc.) have been eradicated. It may even pay to let the site lie as a bare fallow for the summer before planting.

Choose a site near to the location of the animals to which the comfrey will be fed. Clay or loam soil suits comfrey best, whilst sands will require ample farmyard manure (FYM, preferably not stable manure), or pig slurry. Thin soils over chalk can just about support comfrey but thin soils with bedrock near the surface are useless. Start the crop in spring or early autumn with a heavy application of FYM or slurry, plough or dig it in and harrow and roll afterwards.

Comfrey is effectively sterile so it must be established as offsets with a growing point and 4–8 cm (2–3 in) of cut root. These should be planted with the growing point about 0.5 cm ($\frac{1}{4}$ in) below the surface. If large areas are to be planted, it is important that the spacings between the plants are even so that row crop machinery can clean both along the rows and between the plants. In England, 1 m (approximately 3 feet) each way is the usual spacing between rows (in Africa 1.3 m or 4 feet, because they grow faster, in the USA 1 m or 3 feet), and 0.5 m ($1\frac{1}{2}$ feet), between plants. In gardens 0.6 m (2 feet) each way is recommended to facilitate suppression of weeds by the comfrey plants.

In the first year the field will need cleaning both ways to keep the annual weeds down, and mowing over to cut down any flower stems that appear on the young comfrey plants. Unless the crop is going very well, spring plantings should be left until the following spring before cutting. It takes a comfrey field three to four years to build up to its maximum yield, and full production may then be maintained for about twelve years. Then the plants die in the centre as the roots age and the plants can be split and replanted (in spring).

In small stables, leaves may be cut selectively from different plants on a daily basis; but for larger yards and studs, harvesting may be by use of a scythe such as a Turk scythe. On a larger scale, a grass mower with a swath board (see chapter 3) or even up the scale to a forage harvester with choppers and blowers may be used. Cutting may be to about 3 cm above ground level about once a month through the growing season. If the crop 'gets away from you' you can mow it and let the cuttings lie, which has a considerable weed suppressing effect, rather than let it grow too coarse and stemmy. Manure and hoe the crop after cutting.

In temperate climates the first cut is usually in April and the last in October, although in small yards where leaves are cut selectively from individual plants this may be extended by a few weeks. For studs, livery farms and riding schools, it can pay to cut hard in the spring to provide fodder, if you are on clay soils on which the grass is slow to 'get going', whereas on drier, sandier soils where the grass may get going sooner it may make sense to let the comfrey grow on until August/September when grass may be in short supply. As a succulent for stabled horses, cutting throughout the growing season is more likely to fit the rationing system. One person can cut about 300 kg per hour by hand.

Unlike many farm crops, comfrey can take FYM, slurry, poultry manure and even raw sewage although I have reservations about using the latter (see the section on fertilizers below), and tends to require liming about every third winter, depending on soil type. It should be pointed out that as comfrey is not a 'nitrogen fixing' legume, it must have both nitrogen and phosphorus supplied to it; but as it has massive roots it can go down to similar depths to tree roots to obtain potassium and trace elements. Limited information is available about the use of inorganic fertilizers for comfrey.

If you decide you want to get rid of comfrey: (a) 'the best of British luck', (b) try folding or tethering hungry pigs on it to grub up the roots, (c) in the UK you could plough or dig it up in January and February so that the roots are damaged by disturbance in their dormant period, and then plant grass for a five year ley or even permanent pasture, grazing it with sheep, or (d) the most effective comfrey killer is probably ammonium sulphate, but the use of this may affect the choice of crop to be grown afterwards.

This herb is not really suited to grazing by horses, and in any case they seem to prefer to eat it once it has been wilted for a period. Horses will eat up to 18 kg (fresh weight) of comfrey a day, although it would be sensible to give a good allowance of oat straw, chaff or hay to increase dry matter intake. Most stables using comfrey find 2–5 kg per day of the wilted plant is a convenient level. Mr Stephenson fed mares with foals a ration of 6.35 kg (14 lb) fresh comfrey per day 'to prevent foal scour' and forced plants in February under cloches for feeding to early foals.

Further information about feeding comfrey to horses may be obtained from *Horse Feeding and Nutrition* by this author, and about its cultivation from the Henry Doubleday Research Association, details of both of which are given in the reading list and useful address list respectively.

Cultivation of comfrey may be a little out of place in a book about pasture management, but as it doesn't exactly fit in with any 'horsey' subject it tends to be neglected, and any horse keeper who is interested in efficient pasture management should be capable of successfully growing comfrey for horses (and his or her other livestock).

Fertilizers

Many horse paddocks never see an ounce of fertilizer, others see large doses of inorganic fertilizers more suited to dairy cows and causing inordinate management problems to the horse owner. In the absence of detailed knowledge of the specific requirements of horse paddocks, the horse owner must strive to find a happy medium. Regular soil sampling provides a useful aid to pasture management (plate 8).

A definition of the different types of fertilizer would be helpful. Basically *organic* fertilizers are those derived from substances produced by living organisms (e.g. manure, seaweed, etc.), which usually contain carbon compounds, as opposed to inorganic fertilizers which are of mineral origin including, however, metal carbonates, oxides and sulphides of carbon (e.g. limestone, nitrate, phosphate of potash, etc.).

Broadly speaking, the organic fertilizers tend to be chemically more complicated. They provide organic matter to enhance soil structure and

Plate 8 Regular soil sampling provides a useful guide to balanced and economic pasture management. This overgrazed, weed-infested field was brought back as a useful pasture after two seasons of improved management.

may be a source of a wide range of nutrients (e.g. trace elements) in addition to those for which they are specifically being used. They may take a considerable time to break down in the soil, which means they tend to have a slow release effect, and there is less danger of the bulk of the nutrients they contain being washed (leached) out of the soil by heavy rain or flooding.

Conversely, the inorganic fertilizers are usually simple compounds supplying two or three major nutrients, with perhaps some trace elements as impurities, and no organic matter. In general they are quickly released into the soil, whence they may be taken up rapidly by the plant ('luxury uptake'), leached out of the soil or 'locked up' in the soil rendering them unavailable to the crop. This rapid uptake means that great care is needed when using inorganic fertilizers for horse pastures as they can lead to a sudden lush growth (especially nitrates) which can lead to various metabolic disorders due to the sudden change in feed quality. They are, however, useful to obtain rapid growth for fields shut up for hay or silage, although their contribution to trace element levels in the crop may be minimal, a factor which is often neglected.

Organic fertilizers are, on the whole, gentler and safer, giving more even growth and slower release of nutrients, but are a less accurate way of supplying the plant feed. This disadvantage may be reduced if recourse is taken to the manufactured 'organic' 'semi-organic' or so-called 'biochemical' fertilizers which are made. These are of known value like the inorganics, but are slow release and supply organic matter and a wide range of nutrients like the organics, and for grazing pastures at least are probably to be preferred. When comparing costs, the long-term benefits of these products in enhancing soil structure and microbial and worm populations should be considered.

'Artificial' fertilizers are manufactured or refined and the term generally applies to the inorganic fertilizers. 'Compound' fertilizers refer to those formulated to provide more than one major nutrient, for example NPK (nitrogen:phosphate:potash) blends. 'Straights' provide only one major plant nutrient, such as nitrates (e.g. ICI's 'Nitram', etc.).

It seems to me that inorganic nitrates, unless very tactfully handled, are not the most appropriate nutrient sources for horse paddocks. On the other hand, purely organic fertilizers such as FYM may be more idiot-proof, but still have their drawbacks. Cattle manure should be fine, provided you can keep horses off the paddocks whilst it rots down, but it is a rather vague way of supplying nutrients unless you have it analysed. Poultry manure is very nutritious, but may contain harmful additives and in fact, it may have a *very* high nitrogen content which can cause severe scorch on pastures. Pig manure should be avoided because of its high copper content. Human sewage should never be used as it can introduce heavy metals to the soil which may take a hundred years to disappear, having done long-term damage to the micro-organisms which are so very important to soil fertility, apart from any long-term danger to livestock grazing the pasture in the mean time.

The 'happy medium' seems to be the manufactured organic and 'semi-organic' fertilizers which combine to give the best of both worlds.

Why do pastures need fertilizers?

This should be obvious, but to some horse owners it evidently is not. Unfortunately, many horse owners rely on their local vet or riding school proprietor for pasture management advice. Neither group is noted for its abilities in this subject. Studs tend to fare better but can be torn between inappropriate advice from those more used to growing or advising on grazing for cattle, and the cranks and fringe areas which may be associated, perhaps unfortunately, with seaweed and so-forth (more of seaweed later).

Pastures and conservation crops need fertilizer because, in order to grow healthily, the plants take up nutrients from the soil, and these plants are then eaten, or cut and removed. When grazed, some of these nutrients will be returned in the manure, but not always advantageously in the case of horses. In a farm situation, the main nutrients replaced by fertilizers are nitrogen (N_2), phosphorus (P), potassium (K), calcium (Ca) and to a lesser extent sulphur (S), magnesium (Mg) and possibly salt (NaCl). Relatively little attention has been paid to date by the majority of advisers and, therefore, farmers, to trace elements, although there is a growing realisation that these may be deficient in many soils. Levels are to a degree dependent on soil and rock types and urban/industrial pollution. It is quite amazing that this has not been widely realised until recently. Problems due to trace element deficiencies are more likely to occur on sandy soils, e.g. *podzols* where they are scarce anyway and may be leached out, or on highly alkaline (calcareous limestone) soils where they may be 'locked up' in the soil structure, than on say, mineral rich clay soils (not all clay soils are mineral rich).

Intensive grassland for dairy cows or beef cattle may require liming every three to seven years, nitrogen several times a year, plus a spring and autumn dose of phosphate and potash. This system is not appropriate to horse pastures, even intensively grazed ones.

On a fertilizer sack, the three figures refer to the NPK levels in that order, e.g. 20:10:10 contains 20 'units' of nitrogen (N_2), 10 units of phosphate (P_2O_5) and 10 units of potash (K_2O).

So, nutrients may be either naturally present in the soil, and produced by disintegration of mineral rocks by chemical means (releasing trace elements, phosphate and potash), plus the breakdown of plant and animal remains (which also releases nitrogen), or they are added to the soil. Grazing removes more nutrients than are returned as dung and urine, and cutting regimes will remove nutrients altogether. Nutrients may be added actively by plants (e.g. fixation of nitrogen by legumes), or added as organic manures or inorganic fertilizers.

Organic manures

These enrich soil structure by increasing formation of humus, and act as a comparatively 'slow release' and, therefore, safe source of nutrients. Ten tonnes of 'average' farmyard manure contain approximately 30 units of nitrogen, 40 units of phosphate and 75 units of potash. Other nutrients are gradually released as the material is broken down, and the usual application rate to pastures is 25 tonnes/ha (10 tonnes/acre). Ten tonnes of poultry manure supply some 180–300 units of nitrogen, 200 units of phosphate and 80–100 units of potash, and would be used at a lower rate of some 5–10 tonnes/ha (2–4 tonnes/acre).

Farmyard manure is usually spread in autumn or winter. Horse or donkey manure should preferably not be used on horse paddocks, unless *extremely* well rotted, as it will otherwise exacerbate the problem of parasitic worms on the pasture – better if you can sell it to a mushroom grower, or do a swop with a local farmer for cattle manure.

Liquid manures, in particular aqueous ammonia, are not suitable for horse paddocks. Liquid 'seaweed' may be used, but will be dealt with below. Trace elements are likely to be applied as liquids (wettable powders), but will also be dealt with separately.

Other organic fertilizers are blood meal, hoof and bone meal, and meat meal, all fairly rich in nitrogen, but expensive and really the prerogative of the horticulturalist, and used as a nitrogen source. The disadvantages of using human sewage have been discussed above.

Green manuring

This is the practice whereby a fast-growing crop such as oilseed rape is planted in spring, possibly folded with sheep and then ploughed in before sowing a new crop. It could be applicable to ley grass grown for conservation, or before putting in a 'new' permanent pasture, but is not widely practised for grassland.

Artificial inorganic fertilizers

Nitrogen

Nitrogen is important for protein formation and bulky growth of stems and leaves. *Nitrates* are very soluble, so are rapidly taken up by the plant, but are also easily leached (washed out) from soil. *Ammonium salts*, another inorganic source of nitrogen, are less easily leached as they attach themselves to humus and clay particles in the soil, but are also less readily available to the plant – a possible disadvantage to the farmer, but an advantage to the stud manager.

'Straight' nitrogen fertilizers include *sulphate of ammonia* (21% N_2, also called ammonium sulphate), obtained as a natural by-product of coal gas, used mainly in nitrogen-containing compound (NPK) fertilizers, and becoming less common with the move over to 'natural' gas.

Continued use of ammonium sulphate builds up the acidity of the soil, requiring correction by liming. However, this characteristic may be useful if soil alkalinity is a problem. *Ammonium nitrate* (21–23% nitrogen *'lime type'*) provides a rapid release of nitrogen from nitrate and a slow release from the ammonium fraction. Lime is used as a carrier so it tends to correct its own acidity problems. It is most useful on hay crops for horses, once the crop has started to grow. It is very hydroscopic (attracts and absorbs moisture) so must be used up once the bag is opened.

Ammonium nitrate is probably the most common form of nitrogen (e.g. ICI's 'Nitram'), containing 34.5% nitrogen. It has the highest concentration of nitrogen of any normal powder or granulated fertilizer. It can lead to acidity although less so than ammonium sulphate. It may be useful as a top dressing before a hay crop, but is really too concentrated for convenient use on paddocks to be grazed by horses, and its excessive use will lead to lush, sappy growth with a potentially excessive protein content. This is not to say it is not used on horse paddocks, and it can be very useful for eradicating excess clover, but I would suggest the crop should be grazed hard by cattle, or cut for hay (before the clover seeds), before allowing horses back on to it.

Urea (45% nitrogen) may also be used, usually in liquid form for foliar (leaf) application. It is very soluble and although if introduced *very* carefully urea is not harmful to horses, I would again confine its use for crops for conservation rather than grazing.

Gas liquor sodium nitrate and *anhydrous ammonia* are nitrogen sources which I would choose to avoid for horse pastures.

Phosphate

Phosphate encourages root growth and, therefore, affects the whole plant. Deficient plants are stunted and may have a bluish tinge. Many grass crops are deficient in phosphate and horse paddocks may well benefit from regular dressings, subject to soil analysis. As with the nitrates, water soluble phosphates such as *superphosphate* (18–21% water soluble phosphate '*supers*') and *Triplesuperphosphate* (47% water soluble phosphate) are rapidly taken up and are likely to be too 'concentrated' for horse pastures, although use of the former may well be beneficial before and after the grazing season if horses are wintered in. More appropriate to horse paddocks is the citric acid soluble *basic slag* (9–22% phosphate) which is released more slowly and 'gently'. Basic slag is a by-product of the steel industry, from iron foundries, etc. Because the phosphate is released slowly, it can be applied every two to three years, usually being spread in autumn. It also has a liming effect, 100 kg basic slag being equivalent for this purpose to 75 kg chalk. However, due to closure of industrial blast furnaces, there is not much basic slag available now. Most phosphates are obtained from *rock phosphate*, e.g. from North Africa, or Gafsa phosphate (100 kg basic slag is equivalent to 75 kg chalk). *Some* breeders have commented on lowered breeding stock fertility rates after applying basic slag, but I strongly suspect this is due to imbalanced usage or some other problem.

Phosphates are required by the crop to increase seedling growth (and stimulate white clover), and encourage the early ripening of crops. Soluble types can be applied at any time, although autumn and early spring application is particularly suited to grassland for horses. Phosphates are less inclined to be leached out of the soil than nitrates.

Potassium or potash fertilizers

These are mainly derived from clay minerals. Therefore many clay soils are rich in potash, although this is not always the case. However, plants on clay soils may still exhibit potassium deficiency as the clay releases the mineral very slowly. This is more likely to be the case if the land is used intensively for conserved forages (hay/hayage/silage) or arable crops (including oats, barley, maize, etc.) than for less demanding grazing by horses.

Clays and loams obviously contain more potash than sands. There is very little leaching of potash except on light sands, and consequently very little movement of potash in the soil. So, it is important to ensure that it is present where it is needed, i.e. at rooting depth. For this reason, it is important that soil sampling is done at the correct depth for the crop in question, otherwise a false reading may be obtained.

Calcareous (chalky) and peat (organic) soils also tend to be low in potash, although it should be noted that horse manure is rich in potash.

Potassium uptake by the plant is affected by the nitrate and lime levels. Large amounts of these nutrients tend to render the potash unavailable to the plant.

Types of potash

(a) *Muriate of potash* (potassium chlorate), 40–60% potash plus some salt, and *Kanit*, 10–20% potash plus more salt, are more applicable to sugar beet and mangolds (which are maritime plants) because of the salt content.

(b) *Sulphate of potash*, 48–50% potash manufactured from muriate and, therefore, fairly expensive, but stores well as it is less hydroscopic.

(c) *Potassium magnesium sulphate*, 28–40% potash, widely used for grassland for dairy cows because the magnesium content helps to prevent 'grass staggers'. Horses seem to be less sensitive to magnesium levels than dairy cows, often thriving on land where cows would die, although there are indications that some horses are more sensitive to both Mg and K than was previously thought.

(d) *Potassium nitrate*, 10% potash, is more likely to be used in horticulture or as a top dressing for arable crops.

Potassium is most likely to be required on chalk or peat land, soil where high nitrogen levels are added (e.g. dairy pastures) and where crops are removed. Hay is a rich source of potassium in the diet of the horse and obviously if hay crops are regularly taken, replacement of potash in the soil may be required. Potash may also be needed where FYM use is low or of poor quality. Horse manure and urine are rich in potash, but as horses do not graze where they have dunged, unless the field is harrowed frequently to spread droppings, the dunging areas will have more and more potassium and the grazing area can become seriously depleted over a period of time, contributing to the production of 'horse sick' paddocks. Mares are particularly prone to dunging in the same place, so special attention should be paid to the potash nutrition and management of stud paddocks for brood mares. It has been reported by Mary Tavy Archer that 98% of the potassium that has been grazed is excreted in the urine. For this reason, separate soil samples should be taken of grazing and dunging/urinating areas for analysis of potash levels, and selective application of fertilizer should, if possible, be undertaken if this is found to be necessary.

Apart from the worm control aspects, this is another reason why daily collection of droppings is essential to good pasture management. If the area is too large, regular and frequent harrowing is necessary to spread the droppings and thereby spread potassium and encourage break up and drying of droppings and destruction of parasitic worms, and to aerate the pasture. Horses will in fact graze where they have urinated as long as there are no droppings in the immediate vicinity.

Potash fertilizers are twice as effective when 'drilled' rather than broadcast, because this places them in the root zone, so if a paddock is being reseeded it will save fertilizer if a combined drill is used (50 kg drilled is equivalent to 100 kg broadcast potash, or 1 cwt drilled is equivalent to 2 cwt broadcast.

Lime and liming

The application of calcium to pastures is an important factor in crop production. Lime, whose main constituent is calcium, is used basically to increase soil pH, but has the additional benefit of enhancing 'crumb' formation, although this is mainly influenced by soil management. Raising the pH increases humus breakdown as the soil bacteria become more active. (This effect of acidity is why we preserve things in acid, for example onions pickled in vinegar – acetic acid – and grass 'pickled' as silage.) Acid conditions also mean N_2, P and K are not released from the soil, reducing nutrient uptake by the plants, and can favour plant diseases and growth of undesirable grasses and weeds.

Symptoms of lime deficiency include patchy growth and individual plants may be pale (chlorotic) and weak (due to lack of calcium in cell walls). Acid-loving weeds may be present (e.g. sheep's sorrel, spurrey, Yorkshire fog and, in arable crops, corn marigold), and grassland turf will be matted as the dead, unrotted herbage on the soil surface chokes out productive grasses. Deficiencies occur due to leaching (the lime dissolves in the carbonic acid in rain), removal with the crop (hay/hayage/silage/oats/barley/maize), or because it is 'locked up' due to the continual addition of acid-producing substances (e.g. ammonium sulphate).

The degree of soil acidity is measured on the pH scale, which is a measure of the hydrogen ions in the soil. In chemical terms, the pH scale ranges from 0–14; pH 7.0 is neutral, below pH 7.0 is acid, and above pH 7.0 is alkaline. The pH of natural soils in the UK ranges from around pH 4.0 (extremely acid) to pH 8.0 (alkaline, calcareous or lime rich). The majority of UK soils are between pH 5.5 and 7.5. The optimum level for permanent pasture is pH 6.0 on mineral soils and 5.3 on peat soils and where pHs are below this level some form of liming is required. (You can do a rough check using a lime test kit from your local garden centre, but these are not very accurate on organic soils.)

As has been said, the principal effect of soil acidity is to influence the availability of plant nutrients, especially trace elements. Problems attributed to soil acidity are in fact more likely to be due to toxicities of manganese (Mn), aluminium (Al) and iron (Fe), because these elements become noticeably more available for plant uptake at low soil pHs.

Calcium and magnesium are the two main constituents of liming materials which act to correct acidity and raise soil pH. They act in the soil both as bases and essential plant nutrients. (A base is a substance that reacts chemically with acids and neutralises them.) Calcium in soil is held mainly in the clay and organic matter in such a manner that it can be replaced by, or exchanged with, other plant nutrients. When soil is fully charged with this exchangeable calcium carbonate (ECC) it is neutral (i.e. has pH 7.0) and any extra calcium is present as particles of free calcium carbonate (FCC). Soils containing appreciable amounts of FCC are referred to as 'calcareous soils'. Soils developed over chalk and some of the softer limestones and over boulder clays containing chalk are usually rich in FCC (lime) and never need liming. Soils developed on millstone grit, coal measures, triassic sandstones and some of the glacial sands are naturally deficient in ECC and are often acid. As has been mentioned, in some places, even when the parent material (bedrock and subsoil) contains plenty of lime, the overlying soil may be acid.

When lime dressings are added to the soil any deficit of ECC is made up first, and any excess of lime remains as FCC in the soil. The absence of FCC does not always mean that the soil must be limed immediately. Many highly fertile soils are not fully saturated with ECC; although they are slightly acid and have no reserves of FCC they have sufficient ECC to meet all requirements, and need only occasional light dressings of lime.

Measuring the lime requirement

The lime requirement of the soil is the weight of ground limestone or ground chalk required on each hectare of land to raise the soil pH from its present value to the optimum value for the pasture.

The amount of ECC, together with other plant nutrients, which a soil can hold depends upon the soil type and especially its content of clay and organic matter: this is known as its exchange capacity, or cation exchange capacity (CEC). Exchange capacity increases as the amounts of clay and organic matter in the soil increase. *More* lime is required to bring the pH of the soil to its optimum value when its CEC is higher than when it is lower. Thus, to raise soil pH by a given amount, the lime requirement of a clay loam is greater than that of a sandy loam.

Unfortunately, there is no simple relationship between pH values and the amount of lime required. A lime recommendation cannot be given on the basis of pH alone. This is why it is advisable to have your requirement checked for you by ADAS (in the UK) or the State Extension Service (in the USA) or an independent adviser. They will give you the weight of calcium carbonate required to bring the soil pH to the optimum value.

Liming materials

Most naturally occurring liming materials are from sedimentary rocks laid down in one of the following ways:

(a) Accumulation of shells or skeletal remains of animals and plants;
(b) Deposition by marine calcareous algae;
(c) Precipitation of carbonates of calcium or magnesium from water.

Types of lime

The word *lime* has been used to mean calcium oxide (CaO, quicklime), calcium hydroxide ($Ca(OH)_2$, slaked lime) and calcium carbonate ($CaCO_3$, limestone and chalk). The term is also applied to magnesian limestones and burnt limes which contain magnesium as well as calcium. The materials normally used for treating horse paddocks at the present time are:

• ground limestone,
• screened limestone and limestone dust,
• coarse screened limestone and coarse screened limestone dust,
• ground chalk,
• screened chalk,
• calcareous sand – shell sand,
• calcified seaweed,

all of which are sources of calcium carbonate. Most are relatively cheap natural products or by-products. Liming materials are graded according to their neutralising value (NV), which is a measure of their equivalence to calcium oxide. For example, the statement that screened limestone has an NV of 48 indicates that 100 kg of screened limestone have the same effect in neutralising acidity as 48 kg of calcium oxide. The materials for horse paddock treatment are discussed in more detail below:

(a) *Ground limestone.* The NV is 50, which is near the theoretical maximum of 56. It is finely ground limestone or chalk and is the commonest source of calcium carbonate for grassland. The quality depends on fineness of grinding, although if it is too fine it will blow away or clog application machinery. This material is easy to store and not unpleasant to handle.

(b) *Screened limestone and limestone dust.* The NV is about 48. They are by-products of the manufacture of graded stone for road metal, railway ballast and other non-agricultural uses. These dusts should be applied at higher rates than ground limestone, because the coarser particles are not effective in the short term.

(c) *Coarse screened limestone and limestone dust.* The NV is 48. They are similar to the above but coarser, so must be applied at correspondingly higher rates.

(d) *Ground magnesium limestone.* The NV is between 50 and 55. It is more useful for dairy cow paddocks where magnesium deficiency is more likely to be a factor than for horse paddocks, but where magnesium deficiency is known to be a problem this is the product of choice. Two and a half tonnes will supply up to 230 kg magnesium. Where a rapid effect is required for the immediate correction of magnesium deficiency, magnesium fertilizers such as kieserite should be used.

(e) *Ground chalk.* The NV is 50. Considerable quantities of natural chalk are available in some counties in the east and south-east of England. This material has quite a high water content 'as quarried', so must undergo drying before grinding, which can add to the cost of production. Hard chalk quarried further north in England is less moist. In general, the harder chalks should be more finely ground than the soft chalks.

(f) *Screened chalk.* The NV is variable but around 45. It is widely produced in the south and east of England. Whilst the finest grades approximate to ground chalk in efficiency, the coarser grades should be applied at heavier rates to the soil surface and allowed to weather before being harrowed in. In this way, the coarser lumps are broken down to smaller, more efficient particles.

(g) *Calcareous sand – shell sand.* The NV is variable, between 30 and 40. On certain beaches, particularly in north Cornwall and Scotland, the sand contains a high proportion of calcium carbonate in the form of shell fragments. Locally, these are cheap and effective liming materials and are becoming increasingly popular for horse paddocks, particularly as in very large amounts they can lighten soil textures and improve topsoil conditions.

(h) *Calcified seaweed.* The NV quoted by one producer is 56. This product is in fact a type of coral seaweed which is becoming increasingly popular

posed dead grass. If you have some very acid grassland and have decided to plough it up and start again, it is advisable to apply half of the full lime requirement, and harrow it in well before ploughing, to accelerate the rotting down of the old turf. The rest of the lime requirement may then be applied after ploughing.

Following slurry or FYM application, there should be an interval of a few months before liming if loss of nitrogen as ammonia is to be avoided.

Note that large parts of the above section on lime and liming have been taken from the HMSO publication *Lime and Liming*, as has the information for table 6.

Table 6 The pH below which crop growth may be restricted on mineral soils, for a selection of grasses and clovers

Grasses and clovers	pH
Cocksfoot (orchard grass)	5.3
Fescues	4.7
Ryegrass	4.7
Timothy	5.3
Lucerne (alfalfa)	6.2
Sainfoin	6.2
Trefoil	6.1
Vetches	5.9
Clover, alsike	5.7
Clover, red	5.9
Clover, white	5.6
Clover, wild white	4.7

Semi-organic fertilizers

It seems to me that the semi-organic fertilizers lend themselves particularly well to use on horse paddocks. They release their nutrients relatively slowly, are apparently utilised by the plant more efficiently and supply vital trace elements. They are basically made up of inorganic crop nutrients mixed on to a sterile, but not sterilised, organic base (which naturally contains organic matter, live bacteria and trace elements) and is then usually pelleted or granulated for ease of application.

Wastage of nutrients through leaching is minimised when semi-organics are used, and release of nutrients to the soil solution is more closely geared to the uptake demand pattern of the herbage. The risk of 'scorching' through overdosing is reduced, especially in a dry season. Apparently a better balance of nutrient ions in the root area is available, beneficial soil micro-organisms are encouraged by the organic substrate which leads to less soil 'stress'; i.e. the 'land is kept in good heart'. Trace elements may be added into the pellets or granules in a fritted form – not possible with 'prilled' inorganic fertilizers.

It would appear that semi-organics can enhance the palatability of the herbage and reduce sour patches. Because soil micro-organisms are encouraged, earthworm activity is increased, improving aeration, and soil humus content is at least maintained; water retention and, therefore, drought resistance are thereby improved.

Many stud owners are concerned about excessive inorganic nitrate levels, which are thought to cause scouring and other digestive disorders in both foals and adult horses, and even breeding difficulties, and can impart an unpalatable, bitter taste to the herbage. Recent extensive trials have shown that grass grown on Palmers' organic based fertilizers, which contain no added nitrate raw materials, contains only 25% of the nitrates to be found in the grass grown with inorganic NPK fertilizers. As an example, for UK conditions, one manufacturer makes the broad recommendations shown in table 7, although of course you should always be guided by regular soil analyses.

So, lower total levels of semi-organic fertilizers are applied than is the case with inorganic fertilizers because of the more efficient use, and applications are made less often and cause less of a 'flushing' effect because of the slow release. The total cost is roughly the same on a 'per field basis' as for inorganics, if you think solely in terms of the major nutrients applied, but could be considered to have the added cost benefit of the other intrinsic advantages.

Pelleted or granulated semi-organic fertilizers may be direct drilled or machine/hand broadcast with ease. (Recalibration of machine appliances will be needed if they have been used previously for prilled fertilizers.) After application, the fields should be rested for two to three weeks until the fertilizer has broken down and started to work. In very dry weather it may be faster to use the powdered form, although this could blow away!

In the literature, one manufacturer notes in addition to the points raised above that excess nitrogen may increase output, but it also reduces the plant protein *quality*, and in particular the content of the essential amino acid lysine. I feel this point is important as many thoroughbred diets are noticeably deficient in this nutrient. It is also pointed out that

Table 7 Recommendations for semi-organic fertilizer application for studs in the UK

	Analysis of NPK units (%N: %P: %K)	50 kg bags/acre
Stallion studs		
Autumn application Oct/Nov		
Where early spring growth is required	11: 9:10	3
Average soils	3:15:15	3
For soils low in potash	6:12:22	3
Boarding studs		
Spring application		
For potash-rich soils	7:11: 7	3
For low potash soils	10: 5:16	3
For soils with low magnesium index	9: 3:10+3 mg	3

Note
(a) To convert approximately hundredweights (cwt) to kg per hectare, multiply cwt $\times$ 125.
(b) One bag per acre is approximately $2\frac{1}{2}$ bags per hectare (1 ha = 2.471 acres).
(c) To convert % or units of plant food to kg, divide by 2. For example, a product contains 20% N: 10% P: 10% K (units 20:10:10), or 20 units of N, 10 units of P and 10 units of K per 50 kg (cwt); divide by 2, thus each 50 kg (cwt) bag contains 10 kg N, 5 kg P and 5 kg K.

high levels of nitrogen application can lead to 'sod pulling' whereby either tillers or the whole plants are pulled up during grazing, as the high N_2 levels tend to increase leaf growth at the expense of good root development. Too much potash, either from fertilizer or dunging and urination, can inhibit magnesium uptake.

The late Dr André Voisin's 'Law of Minimum' states:

'Insufficiency of an available element in the soil reduces the effectiveness of other elements, particularly manganese and consequently limits vitamin synthesis by the plant. However, excess of an element can also have the same effect on vitamin synthesis and excess sodium and calcium for instance will reduce the carotene content of the plant.'

This latter point is of particular interest for breeding stock as beta carotene, apart from being the precursor of vitamin A, is thought to enhance conception rates and breeding performance in its own right.

The possible need for extra nitrogen should be considered on stud pastures after the leaching effects of a wet winter, whatever the type of fertilizer you are using.

Commercial organic fertilizers

The different farmyard manures, horse and poultry manures, pig slurry, human sewage and green manuring have been discussed above. Other notable sources of organic fertilizer are the various seaweed products. Chilean Natural Nitrate *may* also be used but is not a product I would choose for grassland for horses because of the high sodium content.

Seaweed blends and kelp

Seaweed fertilizers have gained some considerable popularity in the horse world. They tend to have the advantage of a slow release of nutrients because of their organic nature, and they appear to enhance the palatability of the grass and encourage overall grazing. There have been claims that brood mare conception rates and stallion fertility have *improved* (probably due to the trace element content). However, as far as I can see, no controlled scientific trials as to whether the improvement is against *no* fertilizer treatment or a *different* type of treatment have been performed. I would expect to take these claims with a 'pinch of salt' as it may be that *any* fertilizer or trace element treatment would have given this effect.

There are two main problems associated with seaweed products. Firstly, they are invariably liquids and thus require proper spraying equipment, which a horse owner with a small acreage may not possess – a knapsack sprayer may well be adequate for herbicides on a small acreage, and for 'spot' applications, but would not really be a practical proposition for liquid fertilizer application. Thus the help of a contractor or friendly farmer would have to be relied upon. Purchasing old second-hand tractor mounted sprayers is not necessarily a good idea unless you really know what you are looking at; old sprayers can be virtually impossible to calibrate accurately, especially for herbicides where accuracy is vital.

The second disadvantage, which also applies to *feeding* seaweeds, is that the analysis varies constantly with the seasons and time of harvesting/collection, so you can never be quite sure of what you are putting on the land. Also, there is the slight danger that any seaweed collected from where a current passes a sewage outlet, or some type of nuclear or other industrial waste discharge pipe, may be contaminated – perhaps with radioactivity or heavy metals and other undesirable substances. It is, therefore, important that this type of product be purchased from a company which tests the raw material constantly for such contamination and acts accordingly. At least one UK manufacturer's brand literature (Chase – SM3) points out that it uses a blend of seaweeds 'from the *unpolluted* Atlantic coast of Western Ireland' (my italics).

Interestingly, I understand that seaweeds which grow north of the fifty-second parallel are richer in their basic components than the same varieties growing further south. Certainly, as with the semi-organic fertilizers described above, it seems likely that the seaweeds will stimulate the micro-organisms in the soil to help break down organic matter and 'release' otherwise unavailable nutrients for use by the plant.

One of the things which concerns me about the seaweeds is claims that they are 'balanced' fertilizers, or indeed feedstuffs. Balanced for what? I would say they are naturally balanced to be effective seaweeds, not necessarily effective fertilizers or feedstuffs. Perhaps this is 'splitting hairs' but there are enough unjustified claims made for products in the horse world without adding this one to them.

An interesting factor with seaweed fertilizers is that, like other plants, they are naturally rich in chemicals called *cytokinins* (or *phytokinins*) or cell stimulants, which are thought to increase the utilisation of nutrients by the leaves of pasture plants. It has also been claimed that cytokinins enhance plant disease resistance and give several degrees of protection from *marginal* frosts. The cytokinins are also known to delay the ageing or senescence process in plants, thereby prolonging the growth period, and possibly giving a late autumn 'bite' which is particularly useful for animals kept largely at grass.

Although it may be possible to reduce other fertilizer rates (except lime) when seaweed foliar sprays are in regular use, soil analysis should not be neglected as there may be major shortages of nitrogen, phosphate and potash which need to be dealt with, especially on sandy soils.

Some of the leaflets say 'there is no danger of applying too much', but I would seriously query this. There *may* be no danger to the crop, but the overall danger to grazing equines, either of 'too much grass' or of specific nutrient excesses and imbalances, must be considered. I would be particularly concerned about iodine levels. Whilst too little iodine can cause goitre, so can too much, and I have certainly seen goitre in foals whose dams have been *fed* excessive seaweed as a supplement. It said on the packet 'cannot be overfed'! One would hope the price would be a deterrent, but even experienced horse people do some quite hair-raising things sometimes!

A 'typical' analysis, designed to be used at 1 gallon/acre of concentrate (11 litres/hectare) in 20–500 times as much water, is as follows : nitrogen 0.25%, phosphorus 0.26%, potassium 3%, calcium 1.61%, iron 0.0064%, magnesium 0.2%, iodine 0.1%, copper 0.0022%, zinc 0.0019% and sodium 2.05%. Another brand, Maxicrop, is available in the UK with a little added nitrogen (2%), if required. The average analysis of Maxicrop, which is made from a single seaweed species (*Ascophyllum nodosum*), differs significantly from the one given above, but as I have said, all these products are likely to vary somewhat.

On the whole, seaweeds can best be looked upon as a source of trace elements, which will help prevent an absolute deficiency from building up, but are unlikely to be able to 'cure' specific measurable deficiencies. They can be regarded as grass growth promoters. If they are to be used, I would suggest that the soil should first be analysed and any gross deficiencies of NPK, magnesium and calcium and individual trace elements be corrected, *before* going on to an annual or six-monthly regime of seaweed application, checking the soil/herbage analysis again every two to three years.

It is best to apply seaweeds by first grazing or topping the pasture, chain harrowing both ways and then applying the foliar spray. This can be followed by a heavy (slow) flat roll one week later unless compaction is likely to be a problem. I understand that, in the case of Maxicrop at least, if necessary the field can be sprayed and stock put straight back on it. This *will* mean the animals benefit from *that* treatment, but if they can be kept off for between three and twenty-one days, it will give the soil and plants a 'look in' before the animals eat all the seaweed, and will thereby be of longer-term benefit for both soil and plants and ultimately the horses.

The two brands mentioned are available in the USA, Australia, New Zealand, Switzerland, Greece, Spain, Finland, Denmark, Sweden, Sri Lanka, Malaysia, Channel Islands, Guyana, Iceland, Italy, Singapore and other areas.

Maxicrop also make a 'complexed iron' product which can be used if moss is a problem. The grass should be cut or grazed short and the product applied as a 'fine mist spray' on a still day. The moss will blacken and may then be chain harrowed both ways a week later. A second dose may be necessary about one month later in areas with severe infestations. The other nutrients will help the grass grow up and smother the moss out. I do suggest that a pH check be done on fields with this problem. Equines, and other animals, should be removed for a minimum of three days after such treatment.

Tunisian gafsa phosphate

This is not strictly an organic fertilizer, but a naturally occuring one. It is available in the UK as the Timac range of fertilizers. It is a naturally occuring phosphate, available with or without potash, in a number of different 'strengths' or concentrations. This material may be useful for grassland where P and K levels are low. There is also a blend of gafsa phosphate with added potash and calcareous seaweed, to give an analysis of NPK of 0:15:10, plus 35% calcium oxide and 2% magnesium oxide and inherent trace elements.

The product is best applied in autumn, or at least three months before the time of plant growth requirement. It should be thoroughly washed off the herbage by rain before grazing.

I have not gone into detail with regard to desirable soil indices for the various soil nutrients for two reasons. Firstly, these are really a matter for expert interpretation, for which you should consult whoever does your soil analysis, making sure they understand the grass is for mature horses or breeding/youngstock. Secondly, different indices and soil analysis methods are used in different countries, and in the USA differ from state to state, so it would be impossible to correlate data effectively for an international readership in the confines of this book.

Trace elements

When discussing trace elements, you must be clear about whether you are applying them for the health of the grass, the health of the horses eating it, or for both.

The requirements of grassland for trace elements is a much neglected subject. In the past, when land was grazed less intensively and organic fertilizers with a level (if variable) of trace elements in them were applied, acute trace element difficulties were probably uncommon and even subclinical deficiencies would have been less prevalent.

The trace element content of soils varies significantly, certain clay soils tending to have the richest levels. Individual or groups of trace elements may either be significantly deficient or even present in toxic levels (particularly the heavy metals) in different regions. Fall-out from an industrial complex may significantly increase the levels of fluorine, for example, on the sward, and eventually in the soil, sufficient for toxic levels to build up.

Toxic heavy metals (e.g. cadmium (Cd), aluminium (Al) and lead (Pb)) may also appear after deep excavations and other soil disturbances, e.g. building roads, digging pipe lines, even digging deep drains and knocking down old farm buildings/military/industrial establishments can bring up or release heavy metals such as cadmium and lead to clinical or subclinical toxicity levels. I suspect this may also be a factor in predisposing animals to certain unexplained diseases, such as grass sickness, and disorders caused by mycotoxins (see below) and I would always check for disturbances both within a field and upstream/ground water level from it. The fact that individual animals may be affected in a field whilst others are not, *may* relate to grazing preferences and the fact that some plants exhibit a 'luxury uptake' of certain trace elements, which is one of the reasons why (a) some plants grow better on certain soils than others, (b) herbs are richer in minerals than grasses and (c) some plants are poisonous due to the presence of excessive levels of certain trace elements. In fact, the presence or absence of certain grasses, herbs and weeds *can* be an indication of trace element excesses or deficiences likely to be found in a particular field. For example, the absence of clover denotes lack of phosphate, while matted grass, moss, mare's tail, sorrel, buttercups and Yorkshire fog grass denote acidity and trace element imbalances.

Grazing horses are more tolerant of deficiencies, excesses, or imbalances of various trace elements than other farm livestock, partially because they are performing less intensively, although subclinical problems may well be partly responsible for the very low service:conception and conception:foaling ratios which appear to be found acceptable by horse breeders. Levels are likely to be more crucial with the increasing number of endurance horses and COPD (broken wind) affected horses worked hard off grass.

Unfortunately, when trying to evaluate the influence of grazing and other feedstuffs on trace element values in the blood and serum of horses, nutritionists and agronomists are hampered because, astonishingly, research workers who establish veterinary blood and serum parameters very rarely bother to sample and find the levels in the soil/pasture/feed the test animals are exposed to. This seems ludicrous, especially as they tend to add such riders to their research papers as 'the results emphasise the need to establish reference values *related to diet* (my italics) if copper and zinc (or whatever trace element you like) levels in the blood are to be used for interpretive purposes'. This is so obvious that I fail to see the point of spending money on research into blood levels *without* relating it to soil/pasture and feedstuff levels, particularly when money for equine research is so limited. Otherwise, the figures obtained are tantalising but of *very* limited value to the vet, the nutritionist, the agronomist and ultimately the horse! A bit more co-ordination between veterinary and agricultural/feedstuff research facilities is called for here. Most university schools of animal nutrition would be delighted to analyse something useful instead of the same old rat, sheep or cattle diets analysed year after year for teaching purposes.

As the soil–plant–animal pathway for trace elements is necessarily unclear, and soil or herbage analysis results can be misleading, it is often more reliable to treat the animal rather than the soil as far as trace elements are concerned.

There are considerable differences of opinion in agricultural trace element circles (!) about whether trace elements should be applied in chelated or unchelated form. As far as I can judge, the chelated form is probably safer in inexpert hands, and has a rapid 'cosmetic' effect on the grass but is not as effective in the long term. However, it may be useful for urgent 'fire engine' applications where a rapid effect is needed for crop health. Also, there is a possibility that the chelating agents used in such products *may* have an adverse effect on the soil. However, chelates do not 'scorch' the grass when applied, as the sulphate and oxide forms may well do. Non-chelates are applied as wettable powders, and care must be taken not to overdose, as scorching may otherwise occur (quite apart from nutritional imbalances).

Field requirements are assessed by both soil and plant tissue analysis. Selenium, in fact, can be assessed reliably only by plant tissue analysis, but for the other trace elements (and NPK and lime/pH), soil analysis will give a reasonable guide.

As a rough guide, under UK conditions at least, nutrients most likely to be deficient on the Cotswolds and Chalk Downlands are magnesium (Mg), copper (Cu) and iodine (I), whilst on lower land, iodine and *possibly* again copper and magnesium may also be a problem. Alkaline soils tend to lock up manganese (Mn) and copper, so plant tissue levels may bear little relationship to soil levels. If in doubt, both soil and plant tissue should be analysed. Incidentally, iron (Fe) can become toxic at low pHs if regular liming is not carried out, and the highly available iron level can then interfere with copper and manganese uptake and boron (B) availability.

In Britain, from the horse's point of view, there are not yet vast tracts of land exceptionally deficient, or even over-supplied, with specific nutrients, although I believe there are more interrelationships between certain disease conditions in both grass and horses, and soil types than we are as yet fully aware of. However, over forty US states, as well as Northern Ontario in Canada, have been identified as having soil markedly deficient in selenium (Se), a tendency with acid soils. In fact, in large areas of New Zealand, raising of stock was practically impossible before the advent of selenium supplementation. Forty-seven per cent of farms surveyed in 1979 by the MAFF throughout Britain, particularly in hill areas, were considered unable to provide sufficient selenium for grazing livestock. Problems in horses are possible in the Scottish Highlands and Borders, Welsh Hills, Shropshire and north Cornwall where there are areas of acid soils, high rainfall and a basis of 'old hard rocks'. However, in the USA, there are areas of the Rocky Mountains and 'Great Plains States' with alkaline soil conditions, where highly *toxic* levels of selenium are present.

Another trace element which may suffer regional variation is iodine, which may be present in excess on pastures and hays in coastal and small island areas, where iodised salt and seaweed supplements may have to be avoided, but may be severely deficient in so called 'goitre belts', such as limestone areas of Derbyshire in north-east England, and California, Colorado, Dakota (North and South), Illinois, Indiana, Iowa, Michigan, Montana, Nebraska, Nevada, New York, Ohio, Oregon, Utah, Washington and Wisconsin in the USA. In these areas, iodised salt should be fed. Both deficiency and excess of iodine can lead to goitre (an enlarged thyroid gland), even in newborn foals whose mothers' diets are unbalanced. Table salt in the UK has added iodine but 'dishwasher' salt does not so should not normally be fed.

It makes sense to keep in touch with area and regional variations in soil type, as well as levels of fertilizer application and grazing or cutting. It also makes sense when feeding to balance pasture, to buy compound feeds (nuts/pellets/coarse mix/sweetfeed) and supplements balanced for the area you live in. In a large country like Australia or North America, buying broad-spectrum vitamin/mineral supplements from the other end of the country does not make a lot of sense and neither does importing certain American and Australian supplements, which have not been specifically rebalanced, into the UK. When it comes to buying grains and forage, you can either decide to grow or buy in locally and balance any known deficiencies or excesses, or buy from elsewhere, preferably somewhere you know to be rich in the nutrients you are short of. Large breeding farms may even have their own agronomists to dictate fertilizer and trace element treatments as appropriate to growers of their forage and cereals.

Even if the soil, and herbage, contain adequate trace element levels, these may still be either in a form inherently unavailable to the horse or locked up by something else in the plant. This is particularly true of *calcium oxalate* which is a chemical appearing in a number of so-called 'tropical grass' varieties. It locks up calcium and phosphorus to a serious degree, in some cases leading to lameness, ill thrift, and swelling of the bones of the head, which may be diagnosed as *osteodystrophia fibrosa* (a form of nutritional secondary hyperparathyroidism).

This problem has been identified in 'improved' pastures in seventeen counties in Queensland, Australia, where animals were grazing oxalate-rich grasses, including buffel grass (*Cenchrus ciliaris*), green panic grass

(*Panicum maximum* var. *trichoglume*), para grass (*Brachiaria mutica*) and others, which are widely used for pasture improvement in some areas because of their drought resistance and persistence under heavy grazing. Similar conditions have also been reported from Brisbane, Australia (where the grass was *Setaria sphacelata* var. *sericea*) and Sri Lanka, the Philippines and Texas (USA). A figure of more than 0.5% calcium oxalate in the grass seems to give rise to problems. Oxalate-containing grasses in South-East Asian countries which have caused problems include *Pennisetum typhoides* and *P. purpureum*.

The condition *can* be fatal, but is generally reversible. It is probably caused by some mechanism interfering with calcium absorption in the intestines; a similar mode of action to calcium phytate (phytin) inhibiting calcium uptake from cereals, especially wheatbran.

Plants seem to contain higher oxalate levels under conditions of active growth, especially during spring and summer, and after a period of above average rainfall. Either the pasture may then contain more oxalate, or it may provide oxalate levels for a longer period of time. So, if you have to graze animals on such pastures, the provision of free access molassed calcium or calcium and phosphorus is more desirable than applying the material as a fertilizer to the soil.

Herbicides

Where significant levels of weeds, whether they be weed grasses or broad-leaved weeds, are present, it may well be necessary to spray them out. Repeated applications may be required where such pernicious weeds as docks, thistles, nettles, buttercups, bracken and ragwort have 'taken hold'. It will also help if the fields are fertilized to help the grasses grow and smother the weakened (sprayed) weeds. Figure 11 illustrates some common grassland weeds.

For the ecologically minded, a new technique is being tried for eradicating bracken, a weed which is apparently strengthening its grip on hill land in the UK by some 3% each year. Scientists at York University are importing *Parathendes angularis* moths from mountainous regions of South Africa, which they hope will breed caterpillars that feed on the weed's stem. The moths are initially being released, after a period of quarantine, onto carefully monitored areas of the North Yorkshire moors and in the Lake District. Prior to their release, it will be necessary to ascertain whether they might also attack rare plants such as Royal Fern. It is to be hoped that this will prove to be a successful and ecologically

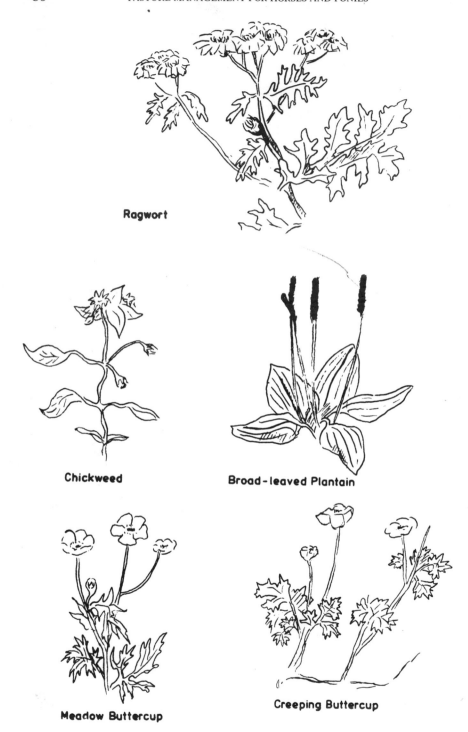

Ragwort

Chickweed

Broad-leaved Plantain

Meadow Buttercup

Creeping Buttercup

Fig. 11 Grassland weeds.

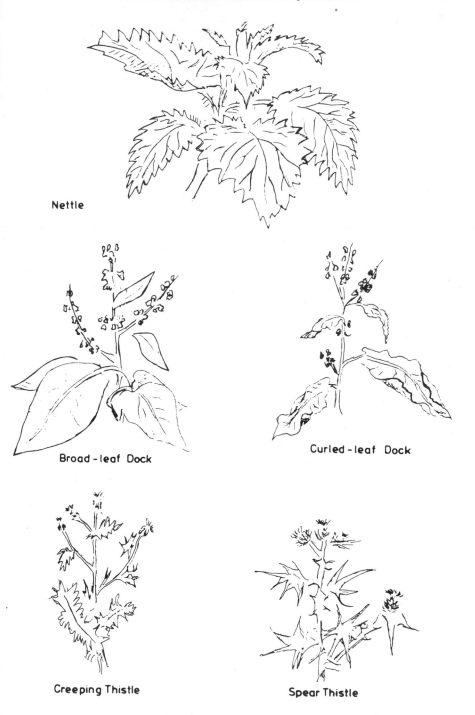

Nettle

Broad - leaf Dock

Curled - leaf Dock

Creeping Thistle

Spear Thistle

Fig. 11 (*contd*)

acceptable method of reducing the levels of this prolific, and indeed, poisonous (see below) weed.

There are two good reasons for eradicating undesirable weeds from grassland:

(a) They may be of poor nutritional value or unpalatable and are thus robbing the soil, desirable herbage, and ultimately the horse, of valuable nutrients and probably smothering out desirable plants as well.

(b) They may be poisonous.

The 'Top Ten' weeds in grassland in temperate regions (spraying periods are for UK conditions) are:

(a) *Nettles* (*Urtica spp.*), best controlled when they are growing vigorously, between the end of April and mid-September. Horses will eat nettles and some relish them, especially if they are cut first and allowed to wilt for twelve hours before feeding, but they should not be allowed to take over a paddock. If you *are* keeping some to feed, cut them before they go to seed.

(b) *Dock* (*Rumex spp.*), curled-leaved dock (*R. crispus*) and broad-leaved dock (*R. obtusifolius*) are both classed as 'injurious weeds' (see below) in the UK. Serious infestations may need to be sprayed in April or May and again in August/September with a repeat spraying the following season.

(c) *Creeping thistles* (*Cirsium arvense*), require spraying when at the 'early bud' stage (June/July). Repeat applications may well be needed on well-established thistles. Creeping thistle (field thistle) is also covered by the *Injurious Weeds Order* (see below).

(d) *Spear thistles* (*C. vulgare*), also covered by the *Injurious Weeds Order*, are dark green in the spring and have long, prickly spines on the leaves which are arranged in a circle at ground level. These should be sprayed in May or June before the flowering stem emerges from the centre of the plant.

(e) *Chickweed* (*Stellaria media*) is an unpalatable, smothering weed. Can be sprayed at any time in the grass crop but preferably when the weed is at the seedling stage. In young leys it is necessary to wait until the grass has begun to tiller before spraying.

(f) *Buttercups* (*Ranunculus spp.*); both creeping buttercup (*R. repens*) and meadow buttercup (*R. acris*) are mildly poisonous if eaten fresh (but safe in hay). They are most common on wet and/or overstocked land. They should be sprayed before flowering, but can be treated later if still growing well. For full benefit, apply fertilizers before spraying to help the pasture smother them out.

(g) *Ragwort* (*Senecio spp.*); if sprays are to be effective, ragwort must be sprayed when it is at the 'rosette' stage before the flowering stem emerges, *and* animals must be kept off the field until it has disintegrated. It may, therefore, be easier and better on small acreages to adopt a policy of pulling up mature plants before seeding and destroying them *out of the field*, as they are palatable and highly poisonous when wilted. Grazing trials in South Africa, reported by the *Equine Behaviour Study Circle Journal*, have shown that some horses graze like 'lawn mowers' eating virtually everything in their path, including young ragwort plants, whilst others may be able to graze more selectively and 'eat round' plants that are not desirable. This may be, in part, an inheritable trait because it is probably associated with shape and mobility of the lips, muzzle and mouth. The South African work was initiated because of the severity of the problem where comparatively large numbers of horses are grazed on range land where it would be impractical and uneconomic to carry out a ragwort eradication programme.

Cutting ragwort just tends to *encourage* more vigorous growth the next year and should never be performed. Ragwort will not be eliminated in one year because of the reservoir of seeds in the soil, but good management to promote a dense and vigorous sward will speed up the process. Ragwort is covered by the *Injurious Weeds Order*.

Although sheep are less susceptible to ragwort poisoning than other species they can still be poisoned and are thus *not* an effective means of eradication.

(h) *Rushes (Juncus spp.)*; first you must identify which rush you are faced with – soft rush or hard rush. Soft rush has a continuous white pith inside its smooth green flowering stem. The hard rush has a bluey green, finely ridged stem which is chambered. The white pith inside is segmented.

(i) Hard rush (*J. inflexus*) should be sprayed with an appropriate material to reduce its vigour, and cut one month after treatment. In the long term, improving drainage and fertilization, including liming, should aid its eradication. If the field is to be reseeded, paraquat can be used to kill the old grass first and it will also kill the hard rush, which has been previously weakened by the earlier treatment mentioned above. Hard rush is poisonous to horses.

(ii) Soft rush (*J. effusus*) is more easily killed than hard rush. For best results, cut the old season's growth, either in the autumn or early spring, and spray the regrowth in June or July before the flowers appear at the end of the stem. In dense infestations, treatment should be repeated for two or three years.

In the UK some measure of protection against ingress of seeds of weeds which are particularly hard to eradicate, or which are poisonous to stock, is given by the *Injurious Weeds Order*, made under the Corn Production (Repeal) Act of 1921 and reconfirmed under the Weeds Act of July 1959.

This order permits the serving on the occupier of any land a notice requiring him or her to cut or destroy any of a number of scheduled weeds; failure to comply with such a notice renders the occupier liable to fines. The weeds scheduled as injurious are spear thistle, creeping thistle, curled dock, broad-leaved dock and ragwort.

Although this order does not often appear to be invoked, neighbours fed up with a 'weed reservoir' when they are trying to keep their own land clear, or local authorities who wish to be very specific about planning permission, rights of way, etc., could well use this order for 'leverage'. (In fact, some local authorities are the worst offenders!) In any case, it makes good economic sense to eradicate these weeds and make the best use of your pasture and remove the very real risk of poisoning from ragwort. It will also help public relations in keeping your fields looking reasonably tidy and not like a waste tip!

Donkeys appear to be partial to thistles but this would be a rather lame excuse for not eradicating these troublesome weeds – or at least cutting them down. (Donkeys probably prefer the fresh, young shoots or cut and wilted mature plants anyway!) Gorse, brambles and other woody weeds can also be troublesome and require a lot of chemical to kill them. An intensive cutting and fertilization programme may be the best management, with or without appropriate spraying.

Poisonous plants

There are a number of grassland and hedgerow weeds which are poisonous to horses and other stock to a varying degree. They are not always avoided in the fresh state, and may become palatable and remain poisonous when wilted or in hay.

Ragwort and bracken appear to be the most serious causes of losses in stock. The former because it is not always avoided by grazing animals and remains poisonous (and becomes palatable) in hay. The latter because, although usually avoided, it is abundant on such an immense acreage of hill land. Of the other poisonous plants found in Britain, yew (*Taxus baccata*) (plates 10 and 11) and laurel (*Prunus laurocerasus*) evergreen shrubs and trees which are commonly planted in places where stock may be able to reach them are probably the most frequent offenders. They are not usually eaten, but during snow, or at other times when herbage is scarce (e.g. the 'typical' horse slum or starvation paddocks of many areas) they may be attractive to horses, ponies and donkeys. Even the dead leaves beneath a yew tree are poisonous. Discarded hedge trimmings or clippings of these shrubs are also dangerous, so if your paddock borders on to gardens, be prepared to have a polite word with neighbours on the subject – if your field *looks* like a dump, it is human nature that people will *use* it as a dump, particularly in urban areas.

Plate 10 Yew (*Taxus baccata*), commonly planted in churchyards in the past, partly because these were not generally accessible to livestock. Highly poisonous.

Plate 11 Yew (*Taxus baccata*) is one of the commonest causes of poisoning in horses. Four times the amount in this photograph could be fatal to a horse – three good mouthfuls? Both leaves and berries are poisonous.

Tables 8 and 10 list the plants found wild, or commonly planted, in Britain and the USA, which are known to be poisonous and which would normally be looked for in investigating any case of suspected poisoning. Many are not weeds in the accepted sense, and some are plants which should not normally be accessible to horses, including garden plants such as box (plate 12). However, it is not unknown for small ponies (and donkeys) to be given the run of their owner's gardens (and houses!) and these are possible candidates for poisoning – as are straying animals with access to gardens, hedgerows, marshes and woodlands.

Much of the information for table 8 has been taken from the book *Agricultural Botany* (second edition) by N. T. Gill and K.C. Vear, with additions from other sources and observations of the author.

An attempt has been made to give some estimate of the plants' toxicity, but it must be remembered that, at best, this can be only a very rough guide. The concentration of poisonous material will vary in different

Table 8 Poisonous plants found in temperate regions (information from various sources)

	Frequency and location	Degree of toxicity	Probable serious or fatal dose (fresh wt)	Poison involved	Whether poisonous in hay or dry state (wilted) (H = harmless, T = toxic)	Clinical symptoms and time before death if fatal	Comments
(a) Grassland weeds *Ranunculaceae:*							
Ranunculus bulbosus (bulbous buttercup)	Very common (dry soils)	Marked	Large	Protoanemonin (a volatile yellow oil)	H	Inflammation and narcosis (narcosis is a condition of profound insensibility which resembles sleep in that the animal can be roused temporarily and is not quite indifferent to severe external stimuli).	Unpalatable, very rarely eaten in significant amounts, but may well be consumed in sufficient amounts to lead to subclinical effects. Also undesirable like all unpalatable weeds because it is robbing the soil and consequently the animal of vital nutrients. Not poisonous in hay.
R. acris (acrid buttercup)	Very common	Slight	Large	Protoanemonin	H	Inflammation and narcosis.	Unpalatable, very rarely eaten in significant amounts, but may well be consumed in sufficient amounts to lead to subclinical effects. Also undesirable like all unpalatable weeds because it is robbing the soil and consequently the animal of vital nutrients. Not poisonous in hay.

	Frequency and location	Degree of toxicity	Probable serious or fatal dose (fresh wt)	Poison involved	Whether poisonous in hay or dry state (wilted)	Clinical symptoms and time before death if fatal	Comments
R. repens (creeping buttercup)	Very common (damp soils)	Very slight	Very large	Protoanemonin	H	Inflammation and narcosis.	Occasionally eaten, rarely serious.
R. sceleratus (celery-leaved buttercup)	Frequent (wet)	Marked	Large	Protoanemonin	H	Inflammation and narcosis.	Very succulent, may be attractive when other herbage scarce.
R. flammula (lesser spearwort)	Frequent (wet)	Marked	Large	Protoanemonin	?	Inflammation and narcosis	Rarely eaten.
R. lingua (greater spearwort)	Rare (wet)	Slight	Large	Protoanemonin	H	Inflammation and narcosis.	Rarely eaten.
Caltha palustris (marsh marigold)	Frequent (wet)	Slight	Large	Protoanemonin	H	Inflammation and narcosis.	Rarely eaten.
Umbelliferae:							
Conium maculatum (hemlock)	Locally common	Very marked	2.5–5.0 kg (5–10 lb)	Coniine and other alkaloids	H*	Narcosis and paralysis; few hours	Poisons evaporate slowly as hay matures therefore matured hay is usually safe. Strong smell, very rarely eaten. Sappy stems may be attractive in dry weather.
Oenanthe crocata (hemlock, water dropwort)	Common (wet)	Very marked	0.5 kg (1 lb) (root)	Oenanthetoxin	T	Narcosis and paralysis; one hour.	Foliage usually avoided, exposed roots may be eaten, e.g. in ditches.

eaten, e.g. in ditches.

Species	Occurrence		Lethal dose	Toxic principle		Symptoms	Remarks
Scrophulariaceae: *Digitalis purpurea* (foxglove)	Very common acid, shade	Very marked	100 g (¼ lb)	Digitoxin and other glycosides	T	Contracted pupils, laboured breathing, convulsions; few hours.	Very rarely eaten fresh. Dangerous in hay.
Compositae: *Senecio jacobeae* (ragwort)	Very common	Very marked and cumulative	1–5 kg (2–10 lb)	Jacobine and other alkaloids	T	Straining, staggering, liver degeneration; up to 1 month.	Dangerous fresh and in hay. Affects all stock, though less common in sheep. Readily invades pasture in a low state of fertility. Horses *normally* avoid it when growing but will readily eat it in hay. The risk is greatest because of the *gradual* destruction of the liver which it causes. May show no effects for some time but eventually lead to debility & death. Dead plants *must* be removed before grazing.
S. aquaticus (marsh ragwort)	Locally common (wet)	Very marked and cumulative	1–5 kg (2–10 lb)	Jacobine and other alkaloids	T	Straining, staggering, liver degeneration; up to 1 month.	Dangerous fresh and in hay. Affects all stock, though less common in sheep. Readily invades pasture in a low state of fertility. Horses *normally* avoid it when growing but will readily eat it in hay. The risk is greatest because of the *gradual* destruction of the liver which it causes. May show no effects for some time but eventually lead to debility & death. Dead plants *must* be removed before grazing.

	Frequency and location	Degree of toxicity	Probable serious or fatal dose (fresh wt)	Poison involved	Whether poisonous in hay or dry state (wilted)	Clinical symptoms and time before death if fatal	Comments
Liliacae:							
Colchicum autumnale (meadow saffron, autumn crocus)	Locally common	Very marked, cumulative	1.5–2.5 kg (3–5 lb)	Colchicine	T	Vomiting, diarrhoea, stupefaction; 1–6 days.	Leaves may be eaten in spring or flowers in autumn. *Very* common on Alpine foot hills but also found in UK. Cannot be controlled in pasture except by laborious handpulling, or digging up bulbs.
Polypodiaceae:							
Pterdium aquilinum (bracken)	Very common (hill land)	Marked cumulative	Large	Thiaminase and possibly other toxins	T	Internal haemorrhages; some days' or weeks	Horses and cattle are affected; it is usually avoided, but so abundant that poisoning is common.
Equisetaceae:							
Equisetium palustre (horse tails, mare's tails)	Locally common (wet)	Marked	Large	Thiaminase and alkaloids	T	Wasting, loss of muscular control	Rarely eaten fresh; very dangerous in hay; other species, also dangerous, thorough cultivation is the only remedy.
Buxaceae:							
Buxus sempervirens (box)	Common in gardens and hedges; locally frequent, grows wild on chalk	Marked	1 kg (2 lb)	? Alkaloids	?T	Purging, congestion of the lungs.	Beware of neighbours dumping hedge clippings!

			0.5 kg (1 lb)				
Laburnum anagyroides (laburnum)	Common in gardens	Very marked		Cytisine (alkaloid)	T	Excitement, convulsions, coma; few hours.	All parts toxic, seeds most dangerous; horses are more susceptible than cattle. Beware overhanging trees.
Rosaceae: *Prunus laurocerasus* (cherry laurel)	Common in gardens but often naturalised	slight–very marked	Varies	Cyanogenetic glycoside	H	Difficult breathing, convulsions; sudden.	Effects variable and may relate to health status of animal. Poisoning not uncommon.
Thymelaeaceae: *Daphne laureola* (spurge laurel)	Rare, woods	Very marked	Small	Glycosides	?T	Purging, intestinal inflammation, narcosis.	Rarely accessible to stock.
D. Mezereum (mezereon)	Rare, woods, frequently gardens	Very marked	Small	Glycosides	T	Purging, intestinal inflammation, narcosis.	Rarely accessible to stock.
Fagaceae: *Quercus spp.* (oak)	Very common	Very slight	Very large	?Tannins	—	Constipation, blood in urine.	Leaves may cause trouble, but acorn poisoning more common (pigs and sheep are immune). Beware overhanging trees and access to woodland.
Ericaceae: *Rhododendron spp.* (rhododendron)	Common in gardens, park land, naturalised acid heath	Very marked	Small	Andromedotoxin and other glycosides	?T	Vomiting, vertigo; death from failure of respiration.	Dangerous to all stock. Beware ponies 'snatching mouthfuls' whilst out riding in woodland.

	Frequency and location	Degree of toxicity	Probable serious or fatal dose (fresh wt)	Poison involved	Whether poisonous in hay or dry state (wilted)	Clinical symptoms and time before death if fatal	Comments
Juncaceae:							
Juncus inflexus (hard rush)	Wet stiff non-acid soil	Marked	?Small	?	?	?	Very tough and apparently unpalatable, but sometimes eaten when pastures bare and losses have occurred. Presence indicates poor or disturbed drainage.
Oleaceae:							
Ligustrum spp. (privet)	Common, especially in garden hedges	Slight	Large	Glycosides	—	Vomiting, diarrhoea.	Rarely serious, but again beware of people dumping clippings.
Taxaceae:							
Taxus baccata (yew)	Common in garden hedges, graveyards and wild in woods in some areas	Very marked	0.5–5 kg (1–10 lb)	Taxine (alkaloid)	T	Irritation and narcosis; sudden.	Probably the most dangerous of all trees – poisoning is frequent. Animals will eat it at all times of the year. Fallen leaves are as poisonous as the fresh plant. Poisoning is usually the result of animals straying or of careless disposal of clippings.

(b) Herbaceous plants of woods, hedges, waste places and gardens

Ranunculaceae:

Anemone nemorosa (wood anemone)	Common in woods	Marked	?Large	Protoanemonin	H	Narcosis and diarrhoea.	Rarely eaten.
Helleborus spp. (hellebores)	Rare in woods, frequent in gardens	Very marked	0.25–0.5 kg ($\frac{1}{2}$–1 lb).	Helleborin and other glycosides.	T	Narcosis and diarrhoea; some weeks.	Rarely eaten, but very dangerous poison.
Aconitum spp. (monkshood)	Very rare in woods, frequent in gardens	Very marked	0.5 kg (1 lb)	Aconitine (alkaloid)	T	Depression and convulsions, paralysis; few hours.	Rarely eaten, but very dangerous poison.
Delphinium spp. (larkspur)	Very rare arable, common in gardens	Very marked	0.5 kg (1 lb)	Alkaloids	—	Depression and convulsions, paralysis; few hours.	Rarely eaten, but very dangerous poison.

Papaveraceae:

Chelidonium majus (greater celandine)	Occasional, waste places	Slight	Large	Alkaloids	—	Vomiting and purging.	Probably rarely serious, but take care when grazing verges and banks.
Papaver rhoeas (poppy)	Very common arable weed, seeds may occur in bought-in cereals	Slight	Large	Rhoeadine	T	Excitement followed by coma.	Rarely important.

Euphorbiaceae:

Mercurialis perennis (dog's mercury)	Very common in woods	Marked cumulative	?Small	Alkaloids	H	Irritation, narcosis.	May be eaten in early spring.

	Frequency and location	Degree of toxicity	Probable serious or fatal dose (fresh wt)	Poison involved	Whether poisonous in hay or dry state (wilted)	Clinical symptoms and time before death if fatal	Comments
Euphorbia spp. (purge (annual))	Common in arable land	Slight	Large	?Euphorbio-steroid	T	Irritation, and inflammation of the mouth.	Danger to stock probably negligible.
Solanaceae:	**Berries of the *Solanaceae* may be harvested accidentally in pea crops and are hard to screen out. Preferably, buy feed peas from known weed free fields, or examine carefully before feeding.**						
Solanum dulcamara (woody nightshade)	Common in hedges	Slight	?Large	Solanine (alkaloid)	?T	Narcosis and diarrhoea.	Rarely eaten.
Atropa belladonna (deadly nightshade)	Local waste places, chalk, shade	Very marked	Varies with animal	Hyoscyamine and other alkaloids	T	Narcosis, dilation of pupils, convulsions.	Berries look rather like black cherries. Very toxic to man, although cattle and horses are less susceptible, but may still suffer from subclinical problems. Roots and berries especially poisonous. (Sheep and rabbits are resistant.)
Hyoscyamus niger (henbane)	Occasional – waste places	Very marked	Small	Hyoscyamine and other alkaloids	T	Narcosis, dilation of pupils, convulsions.	Strong, unpleasant smell, so unlikely to be eaten.
Solanum nigrum (black nightshade)	Common in arable crops so may be bought in in hard feed	Slight	Large	Solanine (alkaloid)	—	Stupefaction, convulsions.	Danger to stock probably subclinical.

Plant	Occurrence	Palatability	Dangerous amount	Toxin	T/H	Symptoms	Remarks
Datura Stramonium (thorn-apple)	Casual, rubbish heaps, etc.	Very marked	Small	Hyoscyamine and other alkaloids	T	Narcosis, dilation of pupils, convulsions.	Probably rarely eaten.
Cruciferae: *Sinapsis arvensis* (charlock (runch, wild radish))	Very common in arable land	Marked (seed only)	—	'Mustard oils'	T	Acute gastroenteritis.	Danger only if fed when in fruit, if seed fed in cake or meal.
Caryophyllaceae. *Agrostemma lithago* (corn cockle)	Often occurs in arable land	Marked (seed only)	0.25 kg (½ lb) seed	Saponin	T	Diarrhoea, wasting.	Danger from feeding cereal 'tailings' containing cockle seed or especially unscreened, home-grown cereals.
Umbelliferae: *Aethusa cynapium* (fool's parsley)	Frequently found in arable crops	Marked	Small	Coniine	H	Purging and narcosis.	Clinical danger to stock probably negligible, although care should be taken with cereal silage.
Primulaceae: *Anagallis arvensis* (pimpernel)	Common in arable crops	Very slight	Large	Glycosides	T	Irritation and narcosis.	Clinical danger to stock probably negligible, although care should be taken with cereal silage.
Gramineae: *Lolium temulentum* (darnel)	Very rare	Marked (seed only)	0.5–1 kg (1–2 lb) (seed)	Temuline (alkaloid)	T	Giddiness, stupefaction.	Formerly common, now of negligible importance, at least in the UK.

	Frequency and location	Degree of toxicity	Probable serious or fatal dose (fresh wt)	Poison involved	Whether poisonous in hay or dry state (wilted)	Clinical symptoms and time before death if fatal	Comments
Cucurbitaceae:							
Bryonia dioica (white bryony)	Locally common in hedges	Marked	2.25 kg (5 lb) root	Bryonin and other glycosides	—	Inflammation and convulsions.	Foliage rarely eaten; dug roots very dangerous.
Dioscoriaceae:							
Tamus communis (black bryony)	Locally common in hedges	Marked	2.25 kg (5 lb) root	Glycosides	—	Vomiting and paralysis.	Roots and fruit dangerous – foliage harmless.
Iridaceae:							
Iris spp. (flags)	Common in marshes, etc.	Marked	—	Iridin	T	Diarrhoea.	Rarely eaten fresh; hay and dug rhizomes dangerous.
Araceae:							
Arum maculatum (wild arum)	Common in waste places and shady hedge banks	?Slight	—	Acrid juice	H	Acute irritation.	Rarely eaten.

Notes:

(1) In most cases only one common name is given as these will often differ in different regions and countries – when cross-referencing, *always* check against Latin name.

(2) Descriptions of plants are beyond the scope of this book and a good, preferably illustrated, botanical handbook should be referred to in case of doubt.

(3) The terms alkaloid and glycoside are explained in the glossary.

(4) Poisonous weeds affecting the USA which are not listed here may be found in Table 10.

(5) H: harmless, T: toxic.

(6) Herbaceous plants are most likely to be a problem when animals stray or have access to garden waste and clippings, or are kept in fields bordered by gardens or woodland. In the latter situation, impress on neighbours the dangers of dumping poisonous 'waste' and keep them friendly by *not* allowing your fields to be full of weeds whose seeds can blow over and infest their gardens!

Plate 12 Box, an ornamental hedge plant which is poisonous to horses. Check for gardeners dumping hedge clippings in your pastures.

parts of the plant and between different individuals of the same species. It will also be affected by the time of year, stage of growth and climatic and soil conditions. Susceptibility of different animals will also be affected by their age, body weight, previous management, state of health, parasite burden/damage and nutritional status and the amount and type of other herbage consumed at the same time. Other factors are the idiosyncrasies of the individual animal, which may or may not be hereditary.

Some poisons may be taken in small amounts without any obvious effects, but when small amounts are taken regularly over a period, they accumulate in the body, possibly exerting a subclinical effect on temperament, disease resistance and 'loss of fitness', until finally the amount present is sufficient to produce clinical symptoms. Such poisons are said to be *cumulative*. Horse tails, or mare's tails, (*Equisetium spp.*) contains thiaminase which destroys vitamin B$_1$, rendering the animals vitamin-deficient (plate 13).

Plate 13 Horse tails or mare's tails (*Equisetium spp.*) is poisonous because it contains an enzyme (thiaminase) which destroys vitamin B$_1$, so the vitamin becomes severely deficient. Common in damp hedgerows and under trees. Rarely eaten fresh, particularly poisonous in hay.

Gill and Vear rightly point out that 'experimental evidence regarding poisonous plants is necessarily difficult and expensive to obtain, and cannot cover a very wide range of conditions; clinical evidence is bound to be scanty in the case of rare plants or plants rarely eaten. The details given are, therefore, often based on old and possibly unreliable reports, or on reports from other countries where conditions may be widely different'. They summarise the position by saying that 'while the presence of poisonous plants in places where they are accessible to stock does not necessarily mean that any animals will ever be poisoned, it does mean that the danger exists'. Insurance companies are likely to take a rather dim view of claims arising from poisoning of stock in *obviously* poisonous weed-infested paddocks.

One species with poisonous varieties which is not mentioned in the tables is ivy (*Helix spp.*). Although I can find no reference to it in either the above book, or *British Poisonous Plants* (HMSO) and several other botanical and veterinary books, it does seem to cause a reaction, probably an allergic condition, if consumed or even 'rubbed' by some horses. This is also the case with nettles (*Urtica spp.*) which can cause a rash called *urticaria*. Unfortunately, a lot of unexplained skin rashes are described as 'nettle rash' ('blane' or 'urticaria') or 'poison ivy' when in fact the cause is unproven. Neither is likely to be fatal, but could cause significant discomfort for the horse and, because of its unsightliness, economic losses through loss of opportunity for the show ring, and rejection of stallions by mare owners and vice versa.

For an insight into the additional poisonous plants encountered in the USA, the following information has been adapted from *Equus* magazine, particularly issues 21, 25 and 41.

According to *Equus*, there are over five hundred varieties of plants toxic enough to kill a horse east of the Great Plains alone! They point out that because of the difficulty of identifying the plants in the field and evaluating specific toxins in the laboratory, few plant-specific antidotes are available. Even in horses which have apparently recovered, lasting and irreversible damage may have occurred to the organs affected by the poison.

'Some of the deadliest substances known are contained in fairly innocuous-looking flora. Many of the identified toxins have been recognised and used – both medicinally and murderously – for centuries, but it is only through modern bioassay techniques that we are beginning to learn the level and extent of the destruction they cause.' (*Equus* no. 21)

One problem is that lay horsemen in different areas will identify the same plants by very different common names, or different plants by similar names. For example, the plants of three genera, *amsinckia*, *senecio* and *crotolaria*, all contain pyrrolizidine alkaloids which are responsible for many equine deaths and reduced performance due to impairment of liver function. North American horsemen call these plants by various names, examples of which are given in table 9. Even in an island as small as Great Britain, common names are numerous and potentially confusing, especially in a crisis like acute poisoning (see table 8).

Table 9 Common names for plants of the *amsinckia, senecio* and *crotolaria* genera

Plant genus	Area	Common name
Senecio	Pacific North-West	Tansy ragwort
	California	Common groundsel
	Texas	Senecio or ragwort
Crotolaria	Mid-west	Rattleweed
	Florida	Showy crotolaria
Amsinckia	California	Fiddle neck

The mode of action of these pyrrolizidic alkaloids is thought to be as follows. When the plant is eaten, the alkaloids are absorbed into the bloodstream and inevitably arrive at the liver, passing into the cells of that organ. Enzymes present in the liver then convert the alkaloids into an active form called *pyrolle*. This substance prevents *mitosis* – the normal division (growth and replacement mechanism) of cells. However, rather than killing them, the affected liver cells and their nuclei continue to grow to up to ten times their normal size. The cells eventually outgrow their own nutrition systems and die.

It is estimated that the liver of the horse can function until 60–70% of it has been destroyed, although the stresses of performance, growth, breeding, lactating, or just poor nutrition may lead to liver failure and death before this extent of damage is reached. The liver may shrink to a quarter of its normal size before failing entirely, although the horse may not appear abnormal, apart from an indefinable loss of performance. Once the liver fails, however, the onset of clinical signs is sudden, and the horse will almost certainly die within three to five days.

The damaged liver allows a build up of waste products, including ammonia, which affect the brain. The horse will show symptoms of brain damage and may stagger, walk round in circles or hang its head and look dopey. Some horses also have yellow mucous membranes and red pigmentation in the urine. Once signs appear, nothing can be done. Even if the horse hasn't eaten enough of the plant to kill him, liver damage (cell enlargement or replacement with fibrous tissue) will inevitably affect the performance of the horse.

I suspect many sluggish or irritable ponies kept on ragwort-infested pastures are in fact affected by low-level poisoning and liver damage. I have seen one extremely valuable and well-known arab stallion with evidence of liver damage (blood enzymes, liver biopsy) who had become very tetchy and continually affected by laminitis, azoturia and unexplained illnesses. In spite of this, his owner still turned him out daily into one of the most heavily ragwort-infested pastures I have ever seen, with the tacit approval of their very well known so-called specialist horse vet! They claimed he didn't graze, but he certainly galloped about – breaking off ragwort plants which would inevitably have wilted and become palatable. I cannot believe he never ate any. It is astonishing that the vet made no comment at all about this practice and felt the liver damage was an 'absolute mystery'.

However, the important thing to note is that subclinical levels of poisoning can and do affect the horse and, through inevitable liver damage, will render it more susceptible to other problems.

Some of the poisonous plants which cause problems in the USA, and are not commonly seen in the UK, include the 'chokecherry', *Prunus virginiana*, which is found lining fence rows in many parts of the USA. It contains a cyanide-forming glycoside, but is not a threat to horses unless they consume wilted or frosted plants. This poisoning is treatable, but usually works too fast for help to arrive in time.

The reason only frosted or wilted plants are poisonous is that an enzyme system in the plant normally prevents the release of cyanide in a form harmful to horses. This is further inhibited by the acid environment of the horse's stomach – unlike ruminants, as the neutral or slightly alkaline environment in the rumen facilitates the conversion of the glycoside into deadly hydrocyanic acid. (The rumen is the 'fermentation chamber' *before* the stomach. Horses ferment feed *after* the stomach.) However, if the plant wilts or freezes, the enzyme system that normally prevents the release of cyanide in the equine breaks down. If this occurs the cyanide is absorbed from the intestine and circulated around the body. It then interferes with the enzyme cytochrome oxidase which is necessary for energy metabolism by the cells, by causing oxygen starvation to every individual cell in the body.

Signs of toxicity reflect this, as the symptoms, which occur minutes after ingestion, will show the horse to be in acute distress – breathing heavily and appearing to be excited or agitated. Soon he will become weak due to the lack of oxygen and may undergo convulsions prior to going into a coma, followed by death. A combination of sodium nitrite and sodium thiosulphate given intravenously may prevent death, but only if given immediately, so is unlikely to be of practical importance.

Other plants which are a particular problem in the USA, especially as horses will actually seek them out to eat them, are white snakeroot (*Eupatorium rugosum*), yellow star thistle (*Centaurea Solstisialis*) and Russian knapweed (*Centaurea picris*).

In the case of the white snakeroot, apparently early European settlers in the American Mid-west found it was responsible for 'milk sickness' caused by human ingestion of milk from poisoned cows – a problem of epidemic proportions in affected areas. This time, the toxin is an alcohol called tremetol, which occurs with a resin acid in the plant. The lethal dose for a horse is anything from 0.9–4.5 kg (2–10 lb). The mode of action is the destruction of heart muscle by enzymes. If the horse is then stressed in any way, heart failure may result, or at the very least reduction of heart function efficiency and deterioration of the liver.

The most noticeable sign of tremetol poisoning is trembling or rapid shaking, especially of the flanks and hindlegs. Once this is evident, death may occur in three to ten days. It is possible to save horses from this type of poisoning by clearing the gut with mineral oil. Recovery is, however, an uncertain proposition largely dependent on the individual horse's stamina. I would suggest the use of probiotics after the mineral oil, to re-establish the gut microbial population. A specially formulated, highly digestible purified diet, avoiding for example excess protein which must be de-aminated by the liver, and possibly even including certain herbs, may well help in this respect.

Yellow star thistle and Russian knapweed cause destruction of areas of the brain which co-ordinate the chewing mechanism. If consumption continues, the poison eventually leads to death by starvation. However, the horse has to ingest 50–150% of bodyweight of the weed before toxicity becomes apparent, and it takes over 275 kg (approximately 600 lb) of the weed (as fed) to cause death. So routine regular field inspections should pick up the presence of this weed before significant damage has occurred. However, if significant quantities have been ingested, although the horse can be kept alive by feeding through a stomach tube or intravenously, he will almost certainly not recover and eventually euthanasia will become inevitable.

Another problem we don't see in the UK is that some horses not only enjoy, but become addicted to species variously called locoweed (*Astragulus spp.*) and crazyweed (as any cowboy film fan will be aware!). Apparently, once they have acquired a taste for it, they will seek it out even when (nutritionally) better forage is freely available. They will happily eat the 275–350 kg (approximately 600–800 lb) needed to produce visible signs of poisoning.

A substance (probably an alkaloid) in the weed permanently damages or destroys individual nerve cells (neurons) of the brain and optic nerve. Once these are destroyed, they cannot be regenerated. Symptoms are the behaviour that give the plant its name, as small objects appear large causing the horse to overcompensate for them. Spatial relationships become distorted and affected animals will walk into fences, quarries, over cliffs, etc. There is no antidote though horses can recover to a degree. Affected horses are a danger to themselves and their handlers and are not safe to ride, even after apparent recovery, as they may 'relapse' at any time, although reproduction seems to be normal and mildly affected mares have been used for breeding provided they do not eat the plant during pregnancy, as otherwise skeletal problems in the foal may result.

Another problem encountered in the USA which we do not see in Europe is the effect of torpedo grass (*Panicum repens*), used widely as a pasture for cattle in the South-east, including Florida. Horses pastured on it develop severe anaemia and die. So far, toxicologists have been unable to isolate the toxic agent or develop an antidote or treatment.

Table 10, taken from *Equus* nos 21 (July 1979) and 25, gives details of some of the poisonous plants which are common in the USA. The ones generally found also in Europe have been indicated (see table 8 for more details).

Whatever part of the world you are in, good pasture management and maintenance of soil fertility will tend to preclude the presence of poisonous plants. However, you should be vigilant, as in most cases the only cure for poisoning is prevention! Be wary also of trees overhanging the pasture which, though out of reach of grazing animals, may drop poisonous fruits or seeds (acorns, conkers, horse chestnuts, etc.) and also a bough in high winds, which as it dies may become palatable and may be poisonous.

Toxicities of plants may vary with soil type and climate, and it may be as well to contact a local agricultural advisor or veterinary school to find out about any plants you should be aware of which are specific to your area.

Examples of trees which *can* safely be used in the vicinity of equines (provided the bark is protected from stripping) and which may be of interest to stud farm owners are Scotch pine, white cedar, willow, white pine, red pine, jack pine, sycamore and tulip tree. Flowering crab and autumn olive have also been suggested, but I am somewhat dubious about these two.

Disease control – mycotoxins

Apart from normal metabolites produced by fungal moulds, which grow naturally on herbage plants, certain strains of particular mould species

Table 10 Poisonous plants commonly found in the USA and their distribution

	Location	Season when most problems occur	Where plant causes most problems	Lethal dose	Clinical signs	Poisonous agent	Treatment	Normally found in UK?
Solanum spp. (Black nightshade (horse nettle))	Throughout USA	Late summer; early fall	Ornamental – often found in pasture	0.5–5 kg	Abdominal pain, inco-ordination, weakness, depression	Solanine alkaloid	No antidote	Yes
Pteridium aquilinum (bracken fern)	NE; NW; Pacific, upper mid-west	Fall, when pasture forage is depleted	Wooded areas	Accumulative, large quantities eaten over 30–60 days or more; exact amount unknown	Thiamine (vitamin B_1) deficiency causes appetite loss, inco-ordination	Thiaminase	Removal from source and massive doses of thiamine will bring about complete recovery if administered before the final stages	Yes
Ricinus communis (castor bean)	California; SW; SE	Any season	Cash crop in California, ornamental elsewhere; seeds main source of toxicity	25–100 g	Mild colic, tumultuous heart beat which shakes entire body	Ricin	No antidote, treat for shock	Yes (ornamental only)
Prunus virginiana (choke cherry (wild black cherry))	SE; NE; mid-west, upper mid-west	When wilting occurs after cutting, frost or heavy storms	Pasture, fence rows	4.5–9 kg	Heavy breathing, agitation, then weakness	Cyanide	Combination of sodium nitrite and sodium thiosulphate if administered soon after ingestion	No

Species	Location	Season	Habitat	Toxic dose	Symptoms	Toxin	Treatment	Recovery
Crotalaria spp. (crotalaria)	SE; Rocky Mountains and plains states	When other forage is poor or unavailable	Pasture, hay	Single ingestion of 10 kg; or lower levels over days or weeks	Liver failure, inco-ordination, depression	Pyrrolizidine alkaloid	No antidote	?
Zigadenus spp. (death camas)	Rocky Mountains; plains states; Pacific NW; California	Spring	Rangeland pasture	Less than 4.5 kg	Salivation, nausea, respiratory difficulty, weakness, depression, coma	Steroidal glycosidal alkaloid	Atropine sulphate, carbachol	No
Equisetum arvense (Equisetum (horse tail, mare's tail))	Upper mid-west, Pacific NW; Rocky Mountains, plains states, California, SW	Any season	Found in wet marsh or meadow areas but poisoning occurs only when plant is consumed in hay	Accumulative; large quantities eaten over 30–60 days or more; exact amount unknown	Thiamine (vitamin B_1) deficiency causes appetite loss; inco-ordination	Thiaminase	Removal of source and massive doses of thiamine will bring about complete recovery if administered before final stages	Yes
Amsinckia spp. (fiddleneck)	California, Rocky Mountains, plains states	When other forage is poor or unavailable	Pasture, hay	Single ingestion of 10 kg or lower levels over days or weeks	Liver failure, inco-ordination, depression	Pyrrolizidine alkaloid	No antidote	?
Cestrum spp. (jessamine)	SE	Any season	Pasture	Accumulative, exact level unknown	Weight loss, poor haircoat, depression	Unidentified	No antidote	?

	Location	Season when most problems occur	Where plant causes most problems	Lethal dose	Clinical signs	Poisonous agent	Treatment	Normally found in UK?
Lanata lanata (Lanata)	SW	When other forage is poor or unavailable	Common ornamental found around houses, highways	10–15 kg	Diarrhoea, hypoglycaemia	Lantadene A	Remove from source and give large quantities of glucose and proteins to support liver and give nutrition	?
Astragulus spp., Oxytropis spp. (locoweed, crazyweed, milkvetch))	SW; Rocky Mountains, plains states	Often comes up in January or February before other forage is available	Rangeland hay, once acquire a taste for it will seek it out	Accumulative, 275–360 kg (or accidental death before)	Inco-ordination, trembling, horse may be dangerous to ride	Thought to be alkaloids	No antidote	No
Nerium oleander, (oleander)	California	Any season	Ornamental house plant clippings	100 g	Aberration of heart beat, profuse diarrhoea	Glycoside	No antidote	Yes, house plants only
Taxus spp., (ornamental yew)	NE; Mid-west	Any season	Ornamental, pasture	Less than 300 g	Subacute: gastroenteritis, Acute: trembling, laboured breathing, collapse	Taxine alkaloid	No antidote	Yes
Conium maculatum (poison hemlock)	Rocky Mountains, plains states	Early spring; root dangerous after this	Low, wet areas of pasture	Small quantities, exact amount unknown	Loss of muscle strength, particularly in hind legs, tremors, coma	Coniine (a volatile alkaloid)	Cathartics, tannic acid, central nervous system stimulants	Yes

Species	Region	When grazed	Use	Amount	Clinical signs	Toxin	Antidote	
Crotalaria spp., (rattleweed, showy crotolaria)	Mid-west	When other forage is poor/unavailable	Ornamental, pasture, hay	Single ingestion of 10 kg, or lower levels over days or weeks	Liver failure, inco-ordination, depression	Pyrrolizidine alkaloid	No antidote	?
Centaurea spp., (Russian knapweed, yellow star thistle)	California, Rocky Mountains, plains states	Any season, growth peak June/July and October/November	Pasture	Accumulative, 275 kg	Brain deterioration, inability to chew/swallow, 'wooden' expression of facial muscles	Unknown	No antidote	No
Senecio spp. (Senecio, (groundsel, ragwort, tansy ragwort))	Rocky Mountains, plains states, Pacific NW; SW; California	When other forage is poor or unavailable	Pasture, hay	Single ingestion of 10 kg or lower levels over days or weeks	Liver failure, inco-ordination, depression	Pyrrolizidine alkaloid	No antidote	Yes
Sorghum vulgare (sorghum (Sudan grass))	SE, SW	Spring, fall, new growth most toxic	Pasture	Accumulative, exact amount unknown	Heavy breathing, agitation, then weakness, dribbling urine, inco-ordination	Cyanide	No antidote	(Yes)
Panicum repens (torpedo grass)	SE	Any season	Pasture	Accumulative, exact amount unknown	Severe anaemia	Unidentified	No antidote	No

	Location	Season when most problems occur	Where plant causes most problems	Lethal dose	Clinical signs	Poisonous agent	Treatment	Normally found in UK?
Cicuta maculata (water hemlock)	Upper mid-west, Rocky Mountains, plains states, Pacific NW	Spring – soft ground, horse pulls toxic roots from ground	Low, wet areas of pasture	50–200 g	Teeth grinding, muscle spasms, respiratory muscle failure	Circutoxin, (a resin)	Remove unabsorbed material from stomach/intestine quickly, sedation may minimise convulsions but if lethal dose has been consumed death is imminent	Yes
Eupatorium rugosum (white snakeroot)	Mid-west	Mid-August through November	Pasture in forested river hills	1–5 kg	Subacute: liver deterioration, inco-ordination Acute: trembles, heart muscle failure	Tremetol	No antidote	No
Robina pseudoacacia (black locust)	E. of Missouri River	Spring, autumn	Hedgerows, ornamental	?	Colic, dilated pupils, weak, diarrhoea, irregular heart beat, depression, death	?	?	No
Aexulus glabra (Ohio buckeye)	—	—	—	—	—	—	—	?
Gymnocladius dioicus (Kentucky coffee tree)	—	—	—	—	—	—	—	?

also produce complex organic compounds that have no obvious function. Many of these metabolic by-products are antibiotic substances, as they inhibit the growth of other micro-organisms. Others are highly poisonous, and are usually called 'mycotoxins'.

Moulds usually reproduce rapidly and freely by the formation of vast numbers of minute airborne spores. Many species are so common in the soil and on rotting vegetation that they are virtually ubiquitous, requiring only sufficient moisture for their spores to germinate and then to colonise any further nutrient material that becomes available. Their growth plays a major role in the natural process of decay. An excess of moisture is usually the key factor in their development on herbage (as well as stored feed). There are a number of moulds which grow on herbage plants. Often they live in relative harmony, but certain moulds, even otherwise benign ones, may produce mycotoxins (myc from *myc*elium – part of the mould growth) under certain conditions.

Fungal spores can survive in conditions that do not permit their growth, and growth can occur in many conditions that do not favour the formation of mycotoxins.

Trace element deficiencies or excesses may increase plant susceptibilities to moulds and/or decrease grazing animals' resistance to mycotoxins. Different herbage varieties may be variably susceptible to infection.

It is characteristic of most mycotoxins that each is produced by a different small group of closely related fungi (sometimes a single species) and only under particular conditions of growth. The mere presence of mould does not, therefore, necessarily imply any significant damage to the pasture, since it may be represented by dormant spores rather than by invasive growth. Even when the mould has grown extensively, and belongs to a species known to produce a mycotoxin, it cannot be concluded that the toxin will be present.

So, unfortunately, mycotoxins are *very* hard to investigate. Fragile moulds may disintegrate, and mycotoxins change their chemical structure due to the physical effects of cutting herbage and transporting it to the laboratory, or changes in the weather, or the microclimate in the herbage, so it is *extremely* hard to prove or disprove that particular animal health problems are in fact related to mycotoxins. (Aflotoxins are mycotoxins appearing in stored grains and oil cakes.)

Mycotoxins can produce varying results in livestock, including abortion. Cases have been documented in cattle ('foggy morning disease') which have been ascribed to mycotoxins produced in muggy or foggy weather conditions. Most mycotoxin-related disorders involve liver or kidney damage/failure, reduced growth rate, infertility, abortion, reduced milk production and even gangrene of the feet (in cattle). There have been no confirmed cases of mycotoxins affecting grazing horses in the UK to date, although it seems likely that mycotoxins (e.g. aflotoxins) in stored feed have caused both acute and chronic problems, as with other livestock.

At the time of writing (mid 1980s), there is some sort of so far unidentified problem affecting the UK, which has killed ten or more *grazing* animals (mostly young ones) and seriously affected at least another ten, especially in the Scottish border country, although there have been cases in Hampshire and Cornwall. The problem was initially diagnosed as azoturia. The animals had muscular spasms, dark brown urine and a lowered glutathione peroxidase activity of the blood – all of which pointed to a selenium deficiency initiated azoturia. Also (in 1984) we had had a very dry summer followed by a very wet autumn, and the lushest autumn grass for many a year, well into November. This lush grass tended to be high in protein content, but very deficient in certain trace elements. However, the affected animals were all grazing animals and youngstock, mostly on a modest level of supplementary feeding with borderline to acceptable selenium intake, although admittedly in areas where selenium deficiency can be a problem. They were not working or on high levels of hard feed, and certainly not in the classic 'Monday morning disease' or tying-up syndrome management (or mismanagement) situation. Appetite, pulse rate, breathing and temperature remained normal.

Interestingly, not all animals in the same paddock were affected, *but* at least one stud manager who lost more than one potentially valuable youngster, noted that stormy nights and muggy mornings tended to precede an 'attack'.

It seems quite possible that patches of mycotoxin (perhaps on specific grass varieties, or areas of slightly different soil type) may very well be implicated. It remains to be seen whether the cause will be detected. If it is mycotoxins, it is hard to see what preventative measures can be taken, other than perhaps the drastic measure of treating all paddocks periodically with fungicide (and then having to keep the stock off them), or keeping stock off grass in 'odd' weather.

It *does* make sense to select fungal disease resistant strains of herbage, although this aspect has received rather less attention from plant breeders, for grazing varieties, than might be desired.

This 'outbreak' has worried UK horse owners, and now I've worried you with it, without offering a solution. However, it is a possibility that should not be ruled out in cases of otherwise inexplicable illness, or 'non-classical' symptoms, or abortion, etc. (the latter *can* certainly be brought on by similar substances – aflotoxins – in concentrated feedstuffs).

In the USA, cases of mycotoxin poisoning of animals pastured on fescue grasses are quite common. Symptoms identified in 1010 mares on 298 farms in Missouri, 26.8% of which were fed on pure fescue (*Festuca*) swards, were agalactia (little or no milk production) in newly-foaled mares (53% affected), gestation periods prolonged by two weeks or more (38% affected), and exceptionally thick placentas (9%), which can suffocate unattended newborn foals (Garrett, Heimann, Pfander and Wilson, 1980). There is also thought to be an association with abortions (18%) and still-births. On the *pure* fescue swards, 16% of the young died. Selenium supplementation appeared to exacerbate some of the symptoms. The symptoms are generally described as 'fescue toxicity', fescue grasses being grown on some 35 million acres in the USA. In fact, it is not the fescue grasses which are 'toxic' but the mycotoxins produced by a parasitic fungus *Epichloe typhina*, an endophyte, which grows in the crown of the fescue plant and spreads up the leaves, especially during the hot, humid summer months.

Epichloe typhina is not visible to the naked eye, and is apparently tasteless to horses. The spread of infection appears to take place only through seeds; an affected plant produces seeds containing the fungus, and new plants from those seeds will also harbour this pernicious endophyte.

Work by Lisa Taylor MS, involving a two-year study of twenty-six pony mares grazing fescue, led to the following conclusions:

'Research with ewes has shown that the mycotoxin depresses the release of prolactin, the hormone necessary for milk production. *Epichloe typhina* also produces chanoclavine I, a chemical known to inhibit mammary development. Levels of cortisol – the hormone responsible for production of the lung mobilisers – were also depressed; because of the mares' deficiencies, several of our foals were alive in the womb, but failed to take a breath at birth' (*Equus* no. 73, p. 102).

Management of fescue pastures for horses involves either reseeding with something else, or planting 'clean' seed chosen from one of several endophyte-free varieties which are available. Seed which has been stored for a year or two will also *probably* be 'clean' but seedling germination is by then likely to be poor. The pasture may be used for non-breeding stock, although there have been reports of laminitis associated with the mycotoxin. It is suggested that hay of another species be offered as supplementary forage. Some workers have found that interseeding (e.g. using a slot-seeder) with clover seems to dilute or even counteract the toxicity of 'fescue fungus'. A 20% clover population is needed so regular re-seeding may be required. Infected fescue should never be allowed to go to seed. Topping it regularly before it seeds will prevent further infection, but will *not* eradicate any existing infection in the sward. Mares should be removed from infected fescue pasture at least 30 days before foaling.

In the USA, fescue plants and seeds can be tested, for a small fee, for the presence of *Epichloe typhina* by Auburn University Fescue Toxicity Diagnostic Center, Department of Botany, Plant Pathology and Microbiology, Auburn University, AL 36849, USA.

Pest control

The following section has been prepared with extensive referral to *Grassland Management* by John Parry and Bill Butterworth.

There are a number of pests which may infest pastures, and especially damage newly reseeded pastures. One authority estimates that 'one in three newly-sown swards suffer considerable, usually irrevocable, pest damage during the establishment phase over at least part of their area'. From time to time the horse keeper may find it necessary to take steps to control these pests. Most pests occur in lowland grass, where losses of up to 40% have been recorded.

In temperate climates, the major pests of grassland, affecting the grass itself rather than the horses grazing it, are slugs, leatherjackets, wireworms and frit fly. The total weight of insects and other invertebrates in grassland can exceed that of the grazing livestock. Because the pests are always there, but rarely seen, the damage and loss of yield often go unnoticed.

Slugs

Slugs attack seedlings both below ground and at the soil surface. The grey field slug (*Derocerus reticulatum*) can severely damage newly-germinated grass on heavy soils in wet seasons. Assess the problem by test baiting with methiocarb pellets covered by a tile or piece of wood. The Ministry recommendation for controlling large numbers of slugs is to broadcast methiocarb or metaldehyde pellets on the soil surface a week before or immediately after sowing. Methiocarb pellets should be applied at the rate of 5.5 kg/ha. The recommended rate of metaldehyde pellets is 31 kg/ha of 3% active ingredient product or half that rate for pellets containing 6% active ingredient.

Slugs are also common in old pasture, and may be controlled by broadcasting 16 kg/ha (or 14 lb/acre) of, for example, 'PP' mini slug pellets immediately damage is noticed. If reseeding a sward, or establishing a new sward, as a precaution always drill 8.0 kg/ha (7 lb/acre) of 'PP' (Plant Protection Ltd) mini slug pellets, for example, along with the grass seed.

Horses *must not* have access to slug pellets, which can be quite palatable but poisonous. I recently heard of a show-jumper who died after munching on a sack of slug pellets left open in an outbuilding.

Leatherjackets and wireworms

Both these pests are common in old pasture, especially in spring and autumn. They both feed on the roots and stems just below the soil surface.

Leatherjacket eggs are laid by crane flies ('daddy long legs') in grassy places where the larvae hatch within ten days to feed either on plant tissues or decaying vegetable matter throughout the winter months. Feeding increases in the spring as they develop and grow. They are particularly active in spring-sown grass. Direct-seeded crops and grassy stubbles are most at risk. Seeds can be hollowed out and destroyed but leatherjackets usually attack tillered plants, biting off the stems at, or just below, ground level. The cut ends of the plants are torn with frayed edges. Leaf tips may be eaten or even pulled below the soil surface. Autumn-sown grass will usually show signs of damage by early April. Spring-sown grass is thinned from April onwards. On heavily-infested grassland, bare patches appear where the plants have been destroyed. These are quickly filled in with weeds or weed grasses.

If ploughing out old grass, it is best performed before early August to remove the egg-laying sites of the crane fly prior to autumn and spring drilling. Chemical control is to spray chlorpyrifos or triazophos at recommended rates.

Wireworms are the shiny golden-brown larvae of click beetles. They are most troublesome following the breaking up of old pasture. The only cultural control method is to plough out grassland in the spring and fallow before sowing the seed in autumn. As with slugs a good, firm seedbed will encourage rapid growth of seedlings and reduce the risk of damage. Until there is a safer effective alternative, the recommended chemical control is to apply 'Gamma-HCH' as a spray at a rate of 1.4 litres/ha of 80% colloidal formulation when an attack is expected, and to thoroughly incorporate during final seedbed preparation.

In existing swards, both pests may be controlled by spraying 1.4 litres/ha (1 pint/acre) of 'Gamma-col' in at least 200 litres/ha (20 gallons/acre) of water. Alternatively you can 'bait' with 16 kg/ha (14 lb/acre) of, for example, 'PP' mini leatherjacket pellets immediately damage is noticed. If you are reseeding an old pasture where leatherjackets were seen in the old sward, before drilling mix 8.0 kg/ha (7 lb/acre) 'PP' mini leatherjacket pellets with the seed when drilling.

Frit fly

Frit fly is more likely to be a problem where ryegrass leys are specifically grown for hay. The larvae of the frit fly burrow into the shoots, killing young plants outright. The larvae are maggot-like grubs 2–4 mm long which spend most of their time in the basal air cavity of the grass tillers. There could be over a thousand larvae chomping away at each square metre of Italian ryegrass, and there are three or even four generations a year! The first damage is likely to occur from May onwards. The second generation emerges in August and September.

Grass species and varieties differ widely in their susceptibility to frit fly. Ryegrass and fescue seem to be more susceptible than cocksfoot and timothy. The debilitating effect of frit fly damage is such that if the pest is controlled, this so-called short-lived grass will flourish for several years! Plant breeders are now selecting far more resistant varieties.

Close grazing by sheep seems to induce much larger populations of frit fly – probably because the larvae find it easier to penetrate the relatively large tiller numbers produced by this grazing. Spring-sown grass crops should be sown before mid-March in southern England and mid-April in the north to reduce the likelihood of damage by first generation maggots. Autumn reseeds should be sown following a gap of at least four weeks from the ploughing or burning off the previous grass crop.

Chemical control is recommended when there is a high risk of frit fly, for example in autumn reseeds, when a pre-emergence application of chlorpyrifos (e.g. Dursban 48E) at a rate of 1.5 litres/ha would be suitable. With the growing realisation of the extent of insect damage in grassland research circles, new insecticides such as phorate are being considered.

Occasionally old swards or gallops may be subject to an attack by one of these pests, and often the first clue to all but the closest observer is an apparent non-response to fertilizer treatment. It is as well to take expert advice about the treatment of such areas (it is unlikely that a whole field will be affected) and it will be necessary to keep stock off the field until the insecticide has disappeared and the sward recovered. Weed killing and fertilizer application at the same time will hasten the recovery of an existing sward.

Chapter 3
Practical pasture management

Grazing behaviour

As a matter of interest, work in Poland and France observing grazing behaviour in horses has shown that animals kept completely on pasture spend about 36% of their time actually eating, and 23% of their eating is done at night. The hourly feeding patterns showed that sunrise and sunset had an observable effect on feeding activities.

The Polish workers found that grazing time is a heritable trait, progeny from different stallions spending a significantly different period of time grazing. Intake in a 7–10 hour grazing period was found to be around 5.9 kg (13.2 lb), although the size and type of horse was not specified. The herbage species used in the test were, in descending order of palatability, red clover, white bent grass, ladino clover, smooth brome grass, meadow fescue, red fescue and perennial ryegrass. The use of up to 300 kg/ha of nitrogen fertilizer increased the palatability of the herbage, especially for red fescue, couch grass and timothy. It is interesting that red clover was found to be most palatable, as I understand is also the case in the USA, whereas with UK-grown varieties, most are found to be relatively unpalatable to horses.

Ronald Kilgour and Clive Dalton, in their *Livestock Behaviour – A Practical Guide* (published by Methuen in New Zealand and Collins in the UK), point out that 'the normal habit of cropping grass is aided by the mobile and very sensitive upper lip. The upper lip forces grass into the mouth and the incisor teeth nip it off. After a few bites horses move to another area and prefer to graze facing up hill'.

'Foals look awkward when nibbling grass and when very young they compensate for their long legs and short necks by spreading their forelegs astride. As they become older, one foreleg is advanced while the other acts as a prop beneath the thorax or chest.'

Kilgour and Dalton also observe that foals tend to eat faeces within ten days of birth, 'presumably to initiate cellulose digestion in the colon.' Probiotics may also be given for this purpose.

Acreage required per horse or pony

The acreage* required per horse or pony will depend to a great degree on the type and general management of the animal, the soil type and the pasture management capabilities of the owner. Whatever the acreage available, be it half an acre to forty acres, management will be more effective if the field can be split, whether temporarily, with suitable (e.g. reflective) electric fencing, or more permanently. Water should of course be available in each section.

Management of native ponies is easier on a smaller acreage, which can be closely controlled to supply balanced but not excessive herbage. So-called 'starvation paddocks' may 'cure' acute symptoms of laminitis with overfat ponies, whilst prolonging chronic ones, because they address the results rather than the reasons for 'ponies getting fat on fresh air', and should be regarded as an aid to *mis*management. This is a matter of considerable importance.

The acreage required will depend to a certain extent on how much supplementary feed you are prepared to supply, whether animals are to be kept out at grass in winter, and whether you wish to conserve a significant amount of forage. A small pony can be kept quite happily on 0.2–0.4 ha (½–1 acre), *provided* it is suitably fertilized and weeds are controlled. However many horses or ponies you may have, less than 0.1 ha (¼ acre) per horse will mean that the paddocks can only be reasonably regarded as exercise or 'loafing' paddocks or corrals.

For just a few general riding horses or ponies, the ideal would be 0.5–1.5 ha (approximately 1 ½–4 acres) per pony, the whole being split into three or four separate paddocks. This would give reasonable keep for the ponies and enable a paddock to be shut up and rested. Ideally, one paddock (the best drained one for choice) could be sown with an 'early' variety of grass to give an 'early bite' and allowed to recover fully thereafter.

In the most fertile situations, a hectare of grass should supply sufficient grazing and hay for three or four horses (totalling 1200–1600 kg bodyweight) or four to six smaller ponies. Average grassland will maintain approximately two horses per hectare, or summer pasture only for twice that number. Typical unmanaged and unfertilized 'pasture' will support only one horse per hectare.

For studs, the easiest size of paddock to handle is around 2.7 ha (6 acres), although of course much will depend on the topography of the land and the presence of existing hedges, etc. If you are adding fences, try

*1 hectare = 2.4711 acres; 1 acre = 0.4047 hectares (ha)

to follow divisions between soil types to simplify fertilizer usage. It has been suggested that 4 ha (10 acres) per mare is the ideal for a stud, plus 0.4 ha (1 acre) each of a grass or grass/clover ley, or lucerne for hay, and 0.4 ha (1 acre) each for oats or barley if these are to be grown, although in these days of high land costs, I would suggest 1.4–2 ha (3½–5 acres) per mare would be quite acceptable. If you do keep an acreage for hay/oats/ barley production, I suggest you do *not* graze horses on it, in order to keep your home-grown feedstuffs parasite-free. You should also remember that any inherent trace element deficiencies or imbalances in your land are also likely to manifest themselves in the home-produced feed, so you must be doubly sure that you are correcting such problems in your supplementary feeding.

Remember also that if cattle are to be used as a 'management aid', young ones are likely to take more minerals from the soil, or more specifically *return* less via the dung and urine, than older ones, and may also be more rumbustious and likely to cause accidents or damage to fencing.

Grazing management

Given that you now have a reasonably well-drained field, with a sward made up of desirable grass varieties and no injurious weeds, safely fenced and if possible divided up into manageable units using permanent or electric fencing (of the high visibility metal strip type), a management routine can be set up along the following lines.

Horses and ponies on small areas (the backyard horse)

Where a restricted amount of land is being used for the 'family' horse or pony, very close management is required to obtain the best use of the land and keep ponies in optimum condition. The animal's feed requirements must be supplied *in the right quantities* and worm and other parasite control will have to be very closely adhered to.

The first essential is to divide the field so that one part may be rested. I have a friend who, until recently, kept seven ponies and horses on two acres, divided into six paddocks (incorporating a cross-country course in the fences), with a show-jumping field and lunging area. She picked up or spread the dung, killed or pulled weeds, used fertilizer regularly on resting paddocks (spread by hand) and single-handedly produced a set-up that would put most horse owners to shame, *and* kept her land and horses fit and healthy! It *can* be done, but obviously this sort of management will be easier for the owner occupier than the grass livery.

A bare paddock containing only rank inedible grass and poisonous weeds such as ragwort, bracken, mare's tails and buttercups will not provide any keep, and it will drive the animals to eat normally unpalatable poisonous plants – slowly poisoning them and quite probably having a subclinical effect on their temperament, attitude and willingness to work. The use of 'starvation paddocks' for overfat or laminitic ponies, without suitable supplementary feeding, may simply prolong their problems or even exacerbate them, and is *not* a good idea at all. The need is to reduce their carbohydrate intake and this should not be done at the expense of fibre, protein, vitamins and minerals. Try feeding molassed chaff and a feed block and always have a trace-mineralised salt lick available.

Ponies that eat tree bark are probably short of fibre, eating soil may indicate a shortage of salt or minerals and eating dung may indicate disease, or a deficiency of protein or other nutrients, or imbalanced digestive micro-organisms, requiring a probiotic to correct it. Eating soil and bark will also wear down teeth excessively.

Once the field has been divided into manageable units, which will partly depend on the number of animals to be grazed and the cost of fencing, plus supplying water to each unit, it is advisable to have the soil analysed on a routine basis, which may be done free by a fertilizer company, (see plate 8) or for a charge by ADAS (UK) or your State Extension Service (USA). Make sure they understand that the field is for grazing horses, and whether or not a hay cut is planned. This analysis should tell you how much lime to use (if any), and whether N (nitrogen), P (phosphate) and K (potash) are required.

Unless you are grazing very intensively, you are unlikely to need to add inorganic nitrogen, unless you are trying to eradicate clover. If nitrogen *is* required, I feel it may be preferable and safer for horse paddocks, especially with inexperienced owners, to use an organic nitrogen-containing material for a slower release of nutrients. However, if you intend to shut a field up for conservation as hay or silage, an application of inorganic nitrogen, or N, P and K, may well prove beneficial.

If you regularly spread the droppings by harrowing, K may not be necessary. If you *do* pick up the droppings, or take repeated hay crops, a low level application of K may *then* be required. Much will depend on your soil nutrient status, which you will know from the analysis. Phosphate may well also be required, probably once every second or third year. A suitable liming material should be applied every three to seven years, as necessary, to bring the pH to around 6.5.

Unless you are in an area of known trace element deficiency, or are using only inorganic fertilizers, it is unlikely that you will need to use any trace elements. For the family pony the most convenient and safe way to be reasonably sure of covering all eventualities is to use a suitable liming material, once every three to seven years as necessary. Then, every spring and autumn, more often if necessary, apply a semi-organic NP and possibly K fertilizer (see chapter 2).

If you are going to cut and conserve an area, apply an organic or semi-organic fertilizer (e.g. in the UK, Palmer's Biochemicals, Humber, Sheppey etc.) which will contain a certain amount of trace elements, as well as NPK depending on the specific product selected.

Poisonous weeds such as ragwort (see plate 3) may be pulled by hand and destroyed outside the field, or sprayed when at the 'rosette' stage. The former method is the most satisfactory on small areas. Non-poisonous weeds such as docks and thistles, some varieties of which are 'notifiable weeds' in the UK, can be sprayed when young, for example using a knapsack sprayer bought or hired from a garden centre or agricultural merchant. Alternatively, they may be cut and burned, preferably *before* seeding, if they are too mature to spray. It may take several years to eradicate them from the field.

The increased liming and fertilizer use should gradually kill off any moss and the improved quality sward should choke out undesirable grasses and weeds such as couch grass (*Agropyron repens*) and buttercups. Stubborn cases may require spraying, broadcasting of new grass seed, a heavy fertilizer application, followed by cutting or, if possible, heavy grazing by cattle or sheep, to smother out the weeds and get the new, desirable grass varieties growing vigorously.

Bare patches, especially around gates and water troughs, can have some sort of hardy, prostrate-growing grass seed scattered in spring and autumn (e.g. *Agrostis*), to provide extra grass and help to stabilise the area. If animals are to be kept out all winter, especially muddy areas may benefit from a scattering of *non-poisonous* fine hedge clippings, sphagnum moss, or one of the new clay-binding products if available.

The sward should, if possible, be 'topped' regularly, and certainly before seeding, to a length of 10–15 cm (4–6 in). At the very least, dunging areas can be hand topped using a Turk scythe to cut off rank grasses before seeding. (The Turk scythe is also useful for cutting down docks, thistles and nettles *before* they seed, but it is *not* suitable for use by children, see below.)

Fields should be harrowed weekly if the droppings are not being collected, to spread droppings and dessicate parasitic worm larvae and to prevent formation of sour patches in the paddock. Harrowing will also aerate the sward. Even if the droppings are collected, a monthly harrowing to aerate the sward and pull up dead material should be performed.

At its simplest, the field may be harrowed by using a suitably trained pony to drag a twiggy branch or bundle of branches around the field. Better still, use a set of chain harrows, possibly weighted with a log, which may again be pulled by a suitably trained pony or horse and will provide useful exercise. In fact, the act of dragging something helps to build up the *longissimus dorsi* muscles, on which the rider sits on a horse, and will improve the fitness of riding horses, strengthening the backs of 'weak-backed' horses. It also helps to prepare brood mares for the weight of a foal and prevent them from becoming 'sway-backed' as so many elderly brood mares are, and in addition provides a useful training exercise for suitably conditioned youngstock. It also provides amusement for elderly horses retired from work, who may well enjoy pulling a harrow over an acre or two every now again. (*Do not* overdo it with any animal not used to the type of work.) Alternatively, the harrow may be pulled by a car, Land Rover (or other four-wheel drive vehicle) or a tractor. (Fig. 21.)

It is important to note that moss-infested fields should *not* be harrowed, as this will spread the moss. First treat it with lime and iron oxide to kill the moss, broadcast herbage seed to fill patches and fertilizer to enable herbage to 'smother' the moss. Once the moss has gone, harrow as above.

Check drainage. Water sources should be clean and, if from a natural source, should have a stony bottom. Otherwise the water is better fenced off, if such problems as 'sand colic' are to be avoided. Stagnant pools are not suitable, and there have been reports of horses contracting *salmonella* infections after drinking from pools frequented by ducks.

Water troughs are best sited on a well-drained site, and may be placed across a fence so that they supply two fields at once, or along the fence with a bar above the trough, high enough to discourage jumping from one paddock to the next. The former method has the disadvantage that it forms a potentially dangerous projection in the field, especially for youngstock, and the latter is an invitation to some horses to jump out or bang their heads on the bar!

Make sure foals and small ponies can see into the trough, and cannot roll under it, and that it is checked daily to be in full working order. Supply pipes should be buried deep enough to avoid freezing in winter. Obviously in cold weather the ice must be broken at least twice a day. In some countries, 'Therma buckets' are now available which are thermally insulated to prevent freezing, or polystyrene floats for the surface of the water, which the horses have to learn to push down in order to drink (there is a danger that such floats may fly away in high winds). I

Plate 14 This typical cattle feeder is not suitable for horses. It has sharp corners, and I have seen a horse which caught its foot in the small 'triangle' above the trough on a wooden version. The horse's foot was almost severed.

understand these are quite successful in preventing icing over of water sources, but would still feel it is necessary to check daily. A heavy plastic (not rubber) ball (too large to be swallowed) bobbing in the trough may also prevent freezing.

Water troughs should be cleaned out regularly. If this is simply not feasible, water snails can be used to keep algae down to acceptable levels, and dead leaves, birds, hedgehogs, etc., removed regularly. It is *not* a good idea to site troughs beneath deciduous trees! Blue-green algae are poisonous and must be eliminated.

Hayracks should be of a safe design, and should be moved regularly to prevent poaching. The rack type allows seeds to drop into the horses' eyes, and I have seen a pony whose foot was almost severed when she

Not suitable

Not suitable

Suitable

Fig. 12. Suitability of field hay feeders for horses.

caught it in the wooden tray that was fitted below the rack to collect spillage, and dragged the whole rack some metres because she could not free herself (plate 14). Some of the round cattle ones are useful (fig. 12), but must be of a type whereby the horse cannot hit its head, or become trapped. Racks should be very heavy so that they cannot fall or be pushed or blown over.

Haynets should be tied *securely* so that they will hang above the level of the top of the leg when empty, otherwise horses or ponies may become entangled in them. As this would be impossible if you have, say, a Shetland pony and a large horse in the same field, mixed groups should be fed on the floor. To avoid fights and bullying, try to supply one more net/heap than the number of horses and ponies, and space them at least four horse lengths apart.

Feeding of hard feed should be supervised, again to prevent bullying. There are a variety of buckets and troughs which can be hooked over a post and rail fence, although I prefer to feed at ground level – which is how the horse has evolved to eat. For this, there is a special non-spill feeder, made of rubber/plastic with four spines so it cannot be tipped over (fig. 13). The ones I have seen are stackable. Alternatively, a bucket may be dropped into a *close fitting* tyre (*not* one which would enable the animal to get his foot trapped in between the tyre and bucket). In fact, there are now custom-made feeders specially designed to fit into a tyre, with an overlapping edge (fig. 14).

The time and effort spent on managing grassland will more than repay you in supplying feed, both grazed and conserved, and improving the health of the horses in your care.

If you are afraid of having too much grass, and your fences are suitable,

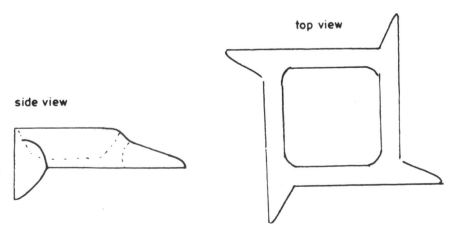

top view

side view

Fig. 13 Non-spill feeder.

Cross-sectional side view

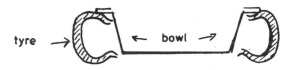

Fig. 14. Using a tyre to prevent feed bowls from over-turning.

you could arrange a beneficial swop with a local farmer, say, allowing a group of bullocks/heifers/sheep to graze in return for hay, or his topping, rolling or harrowing your fields. This will have the added benefit of helping to break the worm cycles in your paddocks. Topping is best performed using a rotary cutter rather than a sickle-bar type as the rotary (e.g. bush hog) mulches the cut grasses more finely and allows them to drop to the ground between the growing grass, whereas the cutter-bar type, whilst necessary for cutting hay, leaves sheaves of grass on top of the new growth and may smother it or slow its growth.

Even if you only use your fields as loafing areas and exercise paddocks, your neighbours will appreciate your keeping weeds under control, otherwise your field may act as a weed seed reservoir and make you *very* unpopular! Trace mineralised salt licks should be available in every field.

Riding schools and other professional facilities

Riding schools, showing yards, dealers, professional show jumpers, etc., will give their clients and sponsors a far better impression of their management abilities if they keep weeds under control and practise reasonable grassland husbandry, and will have healthier, fitter animals into the bargain.

Here the general principles of grassland management apply as for the amateur owner, although obviously things are on a larger scale. It is unlikely to be practical to collect all droppings (apart from smaller riding schools who may have a pool of volunteer labour! – make sure they understand *why* they are doing it) so regular harrowing of both exercise and grazing paddocks is essential. There *is* available a mechanical tractor-mounted unit, which vacuums up droppings from the paddock! Time consuming but faster than a shovel!

Some sort of fertilizer policy, and topping (preferably using a rotary mower) and weed control practice will be required – I am astonished at how many valuable show animals are allowed to graze in paddocks infested with poisonous weeds. Most are not normally eaten fresh, but are palatable when cut and wilted (with the exception of buttercups which are *apparently* not poisonous once they have dried). However, a spirited 'corn fed' show horse galloping about will soon break off some plants with its hooves, and may accidently eat the wilted plant another day. The case of a valuable Arab stallion suffering from chronic liver damage almost certainly due to low-level but continuous ragwort ingestion is described in the previous section.

Obviously, if some forage for conservation can be grown by more efficient grassland management, appreciable savings in feed costs may be made. For an example of this on a large scale, The Donkey Sanctuary at Sidmouth in Devon, England, used to make 4000 bales of hay and buy 30000. After significantly improving their grassland management, they grew 35000 bales and bought none! Obviously, there was a significant cost of production and higher labour requirement, but this must represent a significant saving on feed costs, *and* the donkeys are healthier as they are losing excess weight by controlled access to less grazing!

Producing forage for conservation does require some farming know-how, and hay produced on smaller areas as an 'added bonus' rather than a specific crop is not likely to be of the best quality, so it may be as well to sell it and use the funds thus acquired to buy top quality hay more suited to horse feeding. Indeed, hay grown on horse paddocks may carry a significant burden of worm eggs so may be better sold away, and replaced by hay from paddocks which are never grazed by equines.

It makes sense for a professional 'horse farm' of whatever type in this group to practise a rotational grazing system. It may well pay to set aside certain, better drained, paddocks to provide an 'early bite' in the spring, and they should be sown or top dressed with seed of early-growing grass species. Care should be taken not to overgraze them as the season progresses.

Likewise, a *different* well-drained area may be reserved for late season grazing, using later heading varieties of grass, but it should be borne in mind that a late flush of grass may be of poor nutritional value, and horses should be fed accordingly.

Similarly, spring grass may be excessively rich in protein and soluble carbohydrate (containing more sugar, in fact, than molassed sugar beet pulp!) and low in fibre, and it is prudent to continue to feed hay or chaff until the pasture matures and fibre levels increase. If horses and ponies will not eat hay or straw at this time, molassed chaff may be found to be more attractive to them. Donkeys tend to do better if oat straw is used as they seem to have a more specific requirement for low-feed value fibrous feeds than horses.

Paddocks should be grazed on a rotational basis, giving them adequate time to rest and recover before putting horses back on. If cattle are kept they can be used usefully to 'mop up' the rougher grasses once the horses are removed and help break the worm cycles. The paddocks should then be harrowed and topped, and any fertilizer or herbicides applied. Intensively used paddocks will undoubtedly benefit from a light dressing of fertilizer, say two bags/acre of up to twenty units of nitrogen/acre, more if fields are being shut up for hay. The horses should go back in once the sward is generally 15 cm (6 in) tall. They should not be allowed to eat it down until it is bare as this may kill out the grass, and give weeds and moss (especially in autumn) a chance to gain a foot-hold.

Once a general system of pasture management has been established, modification due to changing weather conditions affecting growth rates, nutrient and trace element availability to the pasture and plant diseases will mean that regular inspections and field walking should be a routine. Actual field operations will never be entirely routine and one should not expect to follow the same system from one year to the next, or even from one field to the next.

Even if fertilizer is not deemed necessary throughout the year, a spring and autumn application of suitable materials, and attention to liming policy, will certainly pay off on the professional unit, both in terms of performance and appearance. Trace mineralised salt licks should be available in every field.

Grazing management on studs

Thoroughbred studs

A stud farm is an intensive grassland unit. The grass may be expected to provide a significant quantity of the animal's nutrient requirements at a time when those requirements are at their very highest – for growth, late pregnancy and early lactation.

The grassland also provides valuable exercise, both for brood mares and excitable youngstock, so apart from meticulous attention to safety of fencing, ditches, fencing gradually projections such as trees (and roots), troughs, etc., and siting of gates, creep areas and so forth, a good springy turf with a tough 'bottom' which will cushion young joints and not be torn up too readily by the hooves of galloping yearlings is essential.

The grass should be nutritious but not lush, and the stud manager should be aware of any potential nutrient deficiencies or imbalances, including trace elements, inherent to the area, which can then be passed on to the grazing animals, and may require specific fertilizer treatment, or a policy of supplying these through the feed. Additional 'hard feed' may be required for most thoroughbred animals, plus some form of long fibre (hay/straw), both in spring when the grass is very lush and during troughs in the growing cycle of the herbage.

Alternatively, it may be decided to rely on the paddocks for exercise and low quality forage, and supply the high protein, energy, vitamin and mineral requirements largely through hard feed. Unfortunately, many studs think they are doing the former and neglect the supplementary feeding, with the resulting poor conception/foaling/lactation/growth and lifetime health results. Many are frightened of 'managing' their grass because they have seen what can go wrong. For example, they've used an agricultural fertilizer regime, producing lush protein and energy-rich grass which is deficient in calcium and trace minerals, leading to joint problems in youngstock – usually put down to 'too much protein'. It is far more likely to be the case on most farms, especially where cereals are the only supplementary feed, that they are just about receiving enough protein for optimum growth, but are badly deficient in calcium and possibly phosphorus, as well as other minerals and trace elements, and even the vitamins required to utilise them properly. 'Too much protein' is an ubiquitous phrase, used frequently to describe problems caused by anything but 'too much protein', and rarely used when problems actually are so caused!

I have in front of me a glossy and initially impressive advertisement showing an aerial view of a thoroughbred stud. The buildings are spotless, the flower beds a well-ordered riot of colour, the gravel drive is beautifully raked, the lawns by the boxes a delight and miles of fencing are painted spotlessly white. Unfortunately, the paddocks are full of weeds, tussocks, bare patches and matted turf indicating a lime deficiency, and are not fit for a bunch of hill cattle, let alone valuable thoroughbred mares, stallions and youngstock. A stud manager should be capable of more than supervising painting fences and polishing pitchforks!

The appalling conception rates of even the better throughbred studs are costing somebody a great deal of money, and it does not take too much imagination to see that improved pasture management, and consequently improved nutrition and environment for mares and youngstock, rather than an over-reliance on expensive drugs and hormones, is likely to pay off in terms of both general health, and specifically improved conception and live-foaling rates. It seems folly indeed not to pay attention to such basically straightforward matters.

It is to be hoped that animals brought up on studs which potentially give them a sound nutritional start in life will eventually command premium prices, particularly as new conditioning techniques seem likely to reveal that we have been relying on genetics alone for top performance for too long. Lifetime balanced feeding and scientifically balanced conditioning *should* give a horse with a reasonable natural ability at least as good a start in life as a fancy pedigree. I hope the 'silly season' in untried yearling prices based on pedigree alone, and usually quite unrelated to earning potential, will soon be brought to a close – for the horses' sakes! The wretched animals are often deemed so valuable/costly that the trainers are afraid to *train* them – 'saving them' for a handful of big races. When you consider that some 70% of racing injuries are caused by fatigue, the large degree of 'natural' wastage in the TB industry, which of course helps the breeders fetch such high prices, does not look so natural. Thankfully, enlightened owners, trainers and even breeders are finally waking up to this, which *must* benefit the horse, and indeed the racing industry, and ultimately the long-term interests of the breeding industry. *Properly* fit horses race sounder, safer, longer, more often and allow the general and fee-paying racing public to have heroes to follow for several years, like the great John Henry, Secretariat and co. At least, in the UK, we have the advantage of steeplechasing, which throws up such equine heroes as the irrepressible Arkle and Red Rum, who year after year captured the imagination of the public.

As I see it the only way racing can continue as an industry in the longer term is to increase the 'gates' and turn it into a sport which is fun for all the family. Increasing the racing longevity of animals is one way of doing this. Correct feeding and grassland management at an early stage is one way of doing that.

Back to grassland management! Assuming again that the drainage has been attended to, a very broad regime of management for a thoroughbred stud in a temperate climate would, subject to soil analysis and local expert advice, be as follows.

Starting in the autumn, subject to analysis, the soil can be limed (every three to seven years; 250 kg–1 tonne/acre; 0.5–2.6 tonnes/ha) and phosphate (40–60 units/acre every two to three years) and potash (30–60 units/acre every year) applied if required. For intensive use some nitrogen will be required, but is best applied in spring or it will be leached away by rain and melting snow over the winter. Where there is a magnesium deficiency and magnesian limestone is not available, some 75 kg of keiserite may be added per tonne of limestone, or suitable magnesium-containing foliar feed may be applied.

Note that pastures should not be grazed until rain has washed inorganic fertilizers at least into the soil. The absolute minimum before grazing can commence is three weeks; seaweeds and other organic fertilizers can be grazed immediately if absolutely necessary. Farmyard manure is best left to overwinter as it tends to render pasture unpalatable.

In spring the pasture should be thoroughly harrowed cross-wise with the long-tined side of a weighted reversible chain harrow to tear out old, dead and tufty grass to open the pasture to the beneficial effects of air, sun and rain, and enable fertilizer to be washed down into the soil and taken up by the plant more quickly. Then, if necessary, it should be dressed with a suitable organic or semi-organic fertilizer and/or trace elements. Any necessary chemical weed control should be undertaken as early as possible, whilst weeds are at the seedling stage and easier to kill with lower doses of chemical. Organic fertilizers, in releasing their nutrients more slowly, reduce the risk of sudden lush growth, although *liquid* seaweeds tend to have the rather cosmetic effect of greening up to a rather sappy pasture, and are better reserved for later in the year, if they are to be used at all. After a week or so, the paddock can be rolled *slowly* again to consolidate the soil and reduce the effects of any poaching, or puffiness due to winter frosts. Avoid too much rolling in autumn.

Thereafter, in spring and summer, the paddocks ideally should be harrowed weekly to spread and break up droppings, preferably on a warm, drying day, as this will dessicate worm larvae, not in damp weather. The 'smooth' side of a reversible chain harrow should be used. Once a month, possibly followed a week later by rolling, unless compaction is a problem, the process may be performed with the tined side, to aerate the turf. Avoid harrowing too much in autumn.

'Top' lush or rank growth to a height of 15–20 cm (6–8 in), to encourage tillering. This should be done once the grass reaches a height of 30 cm (12 in). Herbage plants are at their most palatable and nutritious when they are young and tender. Alternatively, and in fact ideally, follow the youngstock and breeding stock with cattle to 'mop up' the pasture, eat parasitic worm larvae and break their life cycles, and return valuable manure to the paddocks. These cattle can, for example, be sold off fat in August, the paddock rested with a light application of proprietary organic fertilizer ready for the autumn flush of grass growth, horses put on again, followed by a younger bunch of cattle being fed on to the next year, and then resting the paddock over winter. Sheep may be used instead of cattle, but are considered by some to be less effective for mixed grazing with horses and require different fencing. *Do* ensure that any cattle to be introduced are free of brucellosis (*Brucella abortus*), and ringworm, a fungus, as both are contagious to horses, and humans!

Undesirable weeds which have escaped spring eradication can be hand pulled and removed *before* they go to seed, if not too abundant, or they may be topped and sprayed on regrowth, or spot treated using a knapsack sprayer.

Be sure hedge bottoms and surroundings of buildings are not acting as a reservoir of weed seedlings. Those weeds are robbing your grass and your horses of nutrients. Do not kill palatable and beneficial herbs unnecessarily, learn to recognise which grasses and other plants are desirable and which are not – 'a weed is a plant out of place' – and that includes recognising desirable trees, shrubs and hedge plants. A tall tree may be poisonous but inaccessible or unpalatable when growing, but it has been known for leaves to be eaten, once wilted, from branches blown down in gales, even from neighbouring fields, with fatal results.

Trace mineralised salt licks should be available in every paddock.

Warm- and cold-blooded horses

The pasture management on studs for horses other than thoroughbreds will, in principle, differ very little, although it may be less intensive as slower growth and maturation rates for youngstock are necessary, or even desirable. Less nitrogen is likely to be used as protein requirements for slower growth rates are lower, and a pure grass rather than grass/legume sward is desirable.

It is possible that some breeds are not overly efficient at utilising calcium and other minerals in pasture, and care should be taken to ensure the rest of the diet compensates for this. Up to the yearling stage, and for brood mares and stallions, requirements are fairly similar to those of thoroughbreds, although as warm- and cold-bloods tend to be better doers than thoroughbreds, care must be taken to match pasture supplied to requirements.

A smaller acreage may be required for each brood mare, but if animals are being reared on the stud it should be borne in mind that they will tend to be kept for longer than thoroughbreds, so there may be more followers about, up to three years old, whereas the thoroughbred stud will probably only keep stock until they are long yearlings (eighteen months or so), and possibly two year olds. Less hard feed per head will probably be required in addition to pasture.

However, there may be more scope for conservation, and it is suggested that if the stud intends to produce its own hay/silage, specific fields be set aside for this purpose, which are grazed solely by cattle before they are shut up, to reduce the worm burden, and grass varieties and fertilizers chosen accordingly. It should be borne in mind, however, that if the stud has specific nutrient deficiency problems in the soil, these may be reflected in the hay as well as the pasture, and it may be preferable to purchase bought-in forage from another area.

Trace-mineralised salt licks should be present in every paddock.

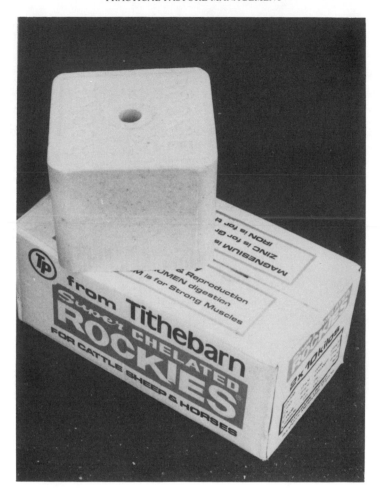

Plate 15 Salt licks should be provided in every pasture.

Ponies, cobs and native ponies

Breeders of thrifty native ponies are likely to have problems with excessively lush pastures, due more to an excess of soluble carbohydrate than protein (lush spring grass often contains more soluble sugar than a mixture of sugar beet pulp and molasses!). Certainly, their grazing should be very tightly controlled, but in such circumstances it is possibly even more important that there are adequate supplies of calcium and other minerals in the soil. Ideally, native ponies are best managed on an extensive upland system, with proper attention to regular worming of both mare and foal. Hard feed may not be deemed necessary, but it is a very good idea to supply 'range blocks', for additional protein (amino

Plate 16 Feed blocks (range blocks in the USA) are an extremely useful way of supporting nutrient levels for pastured horses.

acids), vitamins and minerals, including salt, hung out of reach of sheep if possible, in addition to easily accessible trace-mineralised salt licks.

Recently, a new once-a-week paste vitamin/mineral supplement ('Equi-pep' paste), given orally by syringe, has been developed in the UK by Bayer for working horses. It seems to me that a similar product could be formulated for once-a-week dosing of breeding stock not receiving hard feed, and may be especially useful on upland areas (where it is to be hoped stock are caught and checked over at least once a week). Such treatment would have the added benefit of teaching youngsters to be handled. If the paste cannot be given normally, it can be used to 'butter' a brown-bread and peppermint toothpaste sandwich, which will probably be eaten with relish! Skin implants of slow-release gelatine or liquid glass may also be of use in the future as a means of supplying micronutrients.

The Shetland Pony Society recommends giving fat-soluble vitamin injections every three months, which may be acceptable if ponies are turned away and not caught up for long periods, although it does not, of course, cater for their salt, mineral, amino acid and water soluble vitamin requirements (e.g. folic acid). Another useful means of providing vitamins and minerals to ponies not receiving hard feed, is to give a daily vitamin 'biscuit' such as the German 'Salvana Briquette', available in the UK from Pleyer and Moreton Ltd.

So, native ponies can pose a number of management problems, many of which can be solved by grazing them with cattle or sheep. The main problem is likely to be one of overabundance of unbalanced feed, combined with a deficiency of micronutrients.

Plate 17 The Salvana briquette provides a convenient method of supplementing the vitamin and mineral levels of pastured horses not receiving hard feed. They provide a useful lure for catching up and checking over pastured horses.

Grazing management for donkeys

Donkeys originate from arid parts of the world and do not thrive in wet or muddy conditions. It should be remembered that donkeys have an inherent requirement for low-quality forage and it is suggested that upland grasses, avoiding clovers, suited to sheep pastures should be selected, with an inclusion of herbs in the mixture. It may well be desirable to feed oat- or even barley-straw whilst at grass, particularly during spring, when grass is lush and low in fibre.

Trace mineralised salt licks will be required in every paddock, and feed blocks ('range' blocks in the USA) may be all the additional feeding which is required. Donkeys suffer badly from diseases related to obesity, and it is pointless to spend the winter slimming them down to an acceptable weight and then letting them put it all back on again once the spring grass arrives. Having said this, donkeys should *not* lose weight rapidly, or metabolic disorders may occur. A loss of 1 kg/month is plenty.

Both under- and overgrazing can cause problems and, in general, it is better to understock and keep the paddocks mechanically topped or grazed off with cattle or sheep, or rotated and shut up for hay, than to overstock, as overgrazing tends to encourage low growing weeds. As with horse paddocks, it is desirable to harrow with a pitch pole or heavy, spiked harrow in late February/early March, once the ground is dry enough. Rolling should not be necessary unless the ground is heavily poached, or to roll down stones which may damage forage conservation machinery.

Fertilizer usage will be limited but should not be ignored, particularly in the case of liming. A light spring and/or autumn dressing of a semi-organic fertilizer containing very little nitrogen may be desirable. The Donkey Sanctuary of Sidmouth in Devon, who have around two thousand donkeys under their care, find that they prefer to use a semi-organic type fertilizer as nutrients are released slowly, thus avoiding the danger of a sudden flush of sappy grass growth, and they supply a background level of trace elements. The slow release of nutrients removes the necessity to keep applying fertilizer through the growing season.

For grazing large numbers of donkeys, which are yarded in winter, the Donkey Sanctuary uses the following regime, as described by Paul Svendsen, their farm manager, in *The Professional Handbook of the Donkey* (published in 1986). A semi-organic fertilizer is applied, preferably in late March or before if they can get on the land earlier, before turnout in late April or early May.

'At turnout, the donkeys go into a field that has had slightly less fertilizer and they are strip-grazed behind an electric fence. By using this fencing system, the risk of laminitis is again reduced, as is the risk of the donkeys charging round a large field after six or more months of relative confinement and perhaps getting heart attacks. The two other advantages of this grazing system are that they have fresh grass in front of them every day and they do not dung and trample good grazing land. Using this method, I have grazed 200 donkeys in a ten-acre field for 20 days, and by the time they had finished at the top, the bottom part of the field was ready to be grazed again. With unrestricted grazing, a ten-acre field would have lasted that number of donkeys less than half this time. All the donkeys on Town Barton Farm are weighed every month so that we can keep a careful check on their weight loss or gain. I feel that this is very important for successful donkey management.'

The Donkey Sanctuary are always pleased to advise donkey owners about their care and pasture management, whether for one animal or a hundred, free of charge – although as it is a charitable organisation, donations towards their work are of course welcomed. Contact: The Donkey Sanctuary, Sidmouth, Devon, EX10 0NU. Tel. Sidmouth 6391.

Specific advice on parasite, particularly lungworm, control for donkeys and other animals grazed with them, and a method of eradicating lungworm from a pasture, are given in the section on parasites below (p. 212).

With regard to tethering donkeys, this should be avoided if at all possible. Donkeys should *never* be tethered in extremes of weather (heatwave, cold, strong wind, driving rain, etc.) or isolation, or without free or regular and frequent access to water (which they cannot tip over), or near busy roads or on polluted grass verges. There should be decent forage comfortably in reach and no poisonous plants. The area should be reasonably flat, should not cross a public footpath or bridle way, should not be waterlogged and there should be no large objects within reach that the tether could become entangled with. The Donkey Sanctuary recommend a small radius, say 3 m, for donkeys, necessitating frequent movement and checking by a handler. An area of shade *must* be available, though animals should not be forced to stand under dripping trees and risk rain 'scald'.

If you *must* tether a donkey (or pony for that matter), use a swivel hook in the ground to avoid tangling, and attach the tether to a wide, good quality, stout leather headcollar with a swivel attachment. This should be checked, changed and softened regularly to prevent rubbing. Use of a neck band can be dangerous and should be avoided. A stout but not heavy tether rope (not chain) should be at least 3 m long. Do not leave the donkey unattended until it is well-used to being tethered. If your management requires you to tether a pony or donkey frequently, you should consider whether, in fact, you have the facilities to keep such animals in the first place.

If the proneness to laminitis is your reason for tethering, you should arrange to keep the donkey in a yard or corral instead, with water, feed-grade straw, and a suitable source of vitamins, minerals and protein. Horses, ponies and donkeys under two years of age should not be tethered, and mares or jennies should not be tethered near stallions when in season, or vice versa.

On donkey studs, as with ponies, special care should be taken to ensure that tiny foals cannot squeeze through or roll under fences or become trapped between a water trough and fence, or under a trough, and that they can preferably reach into a trough to drink or have their own water source as soon as they are old enough. Water troughs should be sited either set into, or well away from the fence.

Another important factor to consider when keeping donkeys is that they have very strong 'pair bonding' characteristics and, ideally, should always have a donkey companion. Donkeys kept alone, or who lose or are separated from a companion, can pine or become difficult to handle. Seek advice from The Donkey Sanctuary, who may also be able to lease a suitable companion if your facilities are satisfactory.

The management of mules should follow the basic principles of sound grazing management set out in this book.

Grazing management for other equids – zebra, Przewalski's horses and the wild ass

With the increasing popularity of wildlife and safari-type parks in Europe, and private collections which may include zebra, a few words on their management may well be appropriate.

There are six species of wild equid in the world today, although some, like the Przewalski's horses, are to be found only in zoos. Two species, the Quagga and the European wild horse, became extinct as recently as the middle of the last century. The living species are *Equus przewalskii* (Przewalski's horse); five races of the Asiatic wild ass (the sixth, the Syrian onager, *Equus hemionus* is now extinct), the true Nubian race of the African wild ass, *Equus asinus*, which may not survive in the pure form due to interbreeding with domesticated donkeys; and the zebras. Only the common zebra, *Equus burchelli*, is to be found in large numbers in the wild – other zebras being under the threat of extinction, particularly the Cape Mountain zebra of which there are only at most four hundred left alive.

All these species are represented in zoos, wild life parks and private collections around the world. Their nutrient requirements are likely to be similar to those of the donkey, as they have evolved in harsh conditions, although in captivity they appear to be especially prone to digestive disorders. Certainly, the zebras living wild in some African plains, where soil selenium levels are low, have been diagnosed as suffering from nutritionally-induced muscular dystrophy, and Grevy's zebra is apparently particularly prone to this.

Although the zebras originated in Africa, they generally tolerate the British climate very well. In fact, in Africa, they may have to tolerate a

Plate 18 Pasture management for exotic equids is increasingly relevant for wildlife parks and zoos. (*By kind permission of Mary Chipperfield.*)

range from freezing conditions at night, although the ground may remain warm, to considerable heat (45°C) in the daytime, and may have come from habitats ranging from 4–5000 m above sea level (Kiangs in Tibet), to 60 m below sea level (the Somali Ass of the Danakil depression). However, Grevy's zebra, which comes from nearer the equator, needs to be kept in at night in insulated boxes in Europe. Chapman's zebra can tolerate the cold well, provided it is dry.

Although some species of zebra from the grassland Savannah of East Africa may have relatively high quality feed during the two wet seasons, the average diet contains less than 5% crude protein, and relatively low energy levels. In captivity, stemmy forage may well be preferable and they may be able to graze to a variable degree, for which purpose established permanent pasture containing suitable herbs may be preferred.

Hay of relatively low nutritional value, but clean and mould-free may be selected, and oat-straw would also be a suitable forage. Very little concentrate feeding should be necessary and most brands of horse and pony cubes are likely to be acceptable, along with trace mineralised salt licks of appropriate composition for the areas in which the stock are kept.

Alternatively, a mix containing dried grass/lucerne (alfalfa) meal/ cubes, possibly a little grain and a simple vitamin/mineral supplement may be used. One private park feeds its Grevy's and Chapman zebras 0.66 kg (1½ lb) per day of an oat/wheat bran mix, plus two broad spectrum supplements, which is somewhat excessive. The supplements are Equivite and Selenevite E and in fact only one should be necessary, at a low rate (probably the latter if in a selenium-deficient area, such as much of the hill land in the UK). I would also suggest adding a little limestone, and using chaff instead of bran, or low protein dried grass meal (designated dried green roughage).

The amount of concentrates required is unlikely to exceed 1–2 kg/head per day, depending of course on the quality of the herbage and forage fed, with perhaps a little more for pregnant and lactating females. Alternatively, feed blocks/range blocks, plus forage, could be used. Animals should not be allowed to become overfat.

Regular parasite control should be practised, appropriate to the area. Feed blocks containing an anthelmentic may be used, provided there are no additives, such as *monensin sodium*, which are liable to be harmful to equids. Use only after discussion with the manufacturer, and at your own risk.

Wild equids in captivity are reported to be more subject to digestive disorders than horses or donkeys, and although they have a variable diet in the wild, care should be taken not to vary the diet in captivity, and compound feed suppliers should be chosen who do *not* have a policy of changing compound ingredients without warning. As many of these animals originate in semi-arid regions, they do not tolerate wet weather or mud well, and should be provided with shelters and mud-free standing areas, and field drainage should receive proper attention.

Further information may be obtained by reference to a comprehensive paper called 'The husbandry and veterinary care of wild horses in captivity', by D. M. Jones in the *Equine Veterinary Journal* (1976), volume 8(4), pages 140–46, or by contacting the Royal Zoological Society at Regent's Park Zoo, London, or a veterinarian specialising in the management of exotic species.

Grazing mixed species

Horses and donkeys may be grazed with cattle or sheep as a matter of convenience, or cattle and sheep may be used specifically as a pasture management 'tool' for horse paddocks.

Horses in cattle or sheep pastures

A few points to consider are:

(a) Do not graze horses or ponies with heavily pregnant cows, or especially ewes, as the excitable equines may upset them.

(b) Likewise, do not graze pregnant or newly-foaled mares with excitable bullocks or heifers. Keep an eye on foals with cows, as some cows become quite possessive about a foal, and there are reports of mares trying to 'take over' calves. Similarly, a mare with a cow (or even another horse) as a companion may find the cow driving off the foal when it goes to the mare through jealousy and, again, this has also been reported with mares and calves.

(c) Ensure that horses do not have access to any ruminant (cow/sheep/goat/deer) feedstuffs containing materials likely to be toxic to horses (e.g. the growth promoter *monensin sodium* – 'Romensin' or 'Rumensin'). Horses tolerate urea in feed better than ruminants, but it must be introduced slowly.

(d) If cattle are given free access to a silage face, make sure horses are introduced gradually and prevented from gorging themselves. Apart from the danger of colic to the horses, they can become very possessive about tasty silage and drive off the cattle, so may have to be removed from the field for part of the day.

(e) Check suitability of the fertilization programme for horses. If grass is very lush, restrict grazing.

(f) Check safety and suitability of fences and feeding equipment for horses.

Cattle or sheep in horse pastures

Sheep may be used to help manage horse or donkey pastures, but as they have a close-grazing pattern and apparent preference for shorter grasses similar to horses, they are not as suitable as cattle. Also, sheep fencing (e.g. netting) may not be safe or desirable for horses. *Cattle* are very useful indeed for 'mopping up' excess grass on pastures, and will happily consume the longer rank grasses where horses have staled and dunged, and subsequently avoided grazing. Also, horses will graze right up to cow pats, unlike their own droppings, and the cow dung is a useful manure once spread out by regular harrowing.

The other big advantage, which also applies to sheep in a more limited way, is that cattle do not share the same parasites as horses, and ingestion by cattle of horse parasites can help to interrupt their life cycle and reduce pasture contamination (and vice versa). This, however, should *not* be relied upon as the sole means of parasite control in horses. Some people feel that parasite control is more effective if cattle are put in *after* horses rather than grazing *with* them. Without further research I would say this is debatable though possible.

Ideally, in a stud situation, I would graze horses, then cattle, then top and harrow (and fertilize if necessary), then start again, in view of the potential dangers of grazing foals, pregnant mares or excitable young-stock with cattle, whereas mature horses and ponies can, with care, be grazed alongside cattle, and again the pasture topped and harrowed when they are moved on. Putting horses in after cattle is likely to mean the herbage is 'sweeter' and grazed more efficiently by the horses than in a mixed grazing situation.

Points to consider are:

(a) The *maximum* stocking density will be around 1000 kg livestock per ha under intensive management (for horses) (approximately 900 lb per acre).
(b) Fences should be strong enough to withstand cattle but should not include barbed wire. If grazing is simultaneous, an electric fence wire may be incorporated in post-and-rail fencing to prevent cattle leaning through the fence and breaking it. If grazing is separate and alternated, it may be preferable to run a strip of electric fence about four feet inside the field boundary fence when cattle are in. Apart from the cost of repairing fences, it is obviously not a good idea if cattle break down fences and horses get out.
(c) Young cattle are generally rather boisterous.
(d) Young cattle are also growing fast and some would consider they are robbing the field of minerals and trace elements. However, as both cattle and horses should have an adequate source of balanced minerals, this should not prove to be a problem. In some instances, cattle are more obviously susceptible to trace element deficiencies, imbalances or excesses than horses, probably due to faster growth rates.
(e) Parasites which can be a problem are warble flies, which occasionally attack horses. Ensure cattle grazed with horses are correctly treated with systemic insecticide, if appropriate, to make the field a less attractive environment for the warble flies! The other problem is the skin disorder ringworm. *Never* introduce cattle with ringworm (in fact a fungal infection on the skin and not a worm) into horse pastures, as once you've got it, the animals rub on fences and each other and it spreads like wildfire. There are now feed additives (e.g. Fulcin) which help to eradicate it in infected animals, but obviously prevention by avoidance is better than cure.

In any system of mixed grazing, it is important to ensure that none of the stock is being bullied or worried by others, or being denied access to feed, water or mineral licks.

Pasture renewal/pasture establishment

If land which is not currently being used as pasture is to be put down to grass and herbage for horses, a number of things should be taken into account. If you are thinking of ploughing up existing pasture and reseeding for horses, my advice in 90% of cases is 'don't'! However, if permanent pasture is in a very poor state, and you are not pushed for grazing for a year or so, it really may make sense to plough up and start again. However, if the paddock is 'horse sick' it will need at least four successive arable crops to eliminate the sour patches and start anew. If you are going to take such drastic action, it is best to check the field drainage at the same time.

Obviously the inherent suitability of the land and soil in question for grassland should be considered (especially if the land is not already down to grass), drainage checked and attended to as necessary. You should also take into account the fact that it can take many years to build up a good 'bottom' in a sward to render it resistant to treading, and at the same time provide good springy 'going' underfoot to reduce stress on the limbs. This is especially crucial when considering pasture for valuable youngstock.

Ideally the new sward should first be grazed with cattle or sheep ('the golden hoof'), who should be removed before the new grass is eaten down too tight or the ground becomes muddy and poached, and/or used for silage. At the end of the first year (spring-sown), or better still, autumn of the second year (autumn-sown), horses may be introduced.

Fast-growing ley grasses are not, in general, suited for horse pastures, although it may be of use to put a 'new' field down to a three to five year ley specifically for conservation and, at the end of that period, renovate it by introducing grass varieties more suited to permanent pasture *without* ploughing up, although you cannot hope to achieve the 'best of both worlds'. If you cannot afford to wait, and are thinking of purchasing land for horses, especially youngstock, it may be prudent to avoid land which is not already grassed down, or which has been ploughed up. Having said this, non-pasture land has the benefit that it will not be 'horse-sick' or carry a parasite burden, and you can choose exactly the herbage varieties you wish to use, and manage the sward accordingly.

Unless part of an existing farm set-up, it is most unlikely that a horse enterprise, even a large stud, will possess the appropriate farm machinery and expertise for pasture establishment, and it is likely that a neighbouring farmer, or specialist agricultural contractor, would be brought in to do the work. The principles are the same for any grassland establishment, apart from the choice of a specific herbage seed mixture, although (all things being equal) it is probably more desirable to spray off the weeds and remains of the old crop and direct drill the grass rather than ploughing, from the viewpoint of the desirability of rapidly obtaining a good 'bottom' (and incidentally fuel economy in this energy conscious age).

However, those who object to the use of any herbicides and insecticides may prefer to use ploughing rather than 'minimal cultivations'. A description of 'how to plough' is beyond the scope of this book, but a brief description of the general requirements and sequence of events if you are establishing a new pasture may be useful.

Ploughing

Ideally you would check your lime status, apply half the required lime, then a few weeks later plough thoroughly, burying *all* the trash, and apply the rest of the lime. Then plough furrows can be left to overwinter, and once a tilth has been obtained, the desired seed mixture, chosen to suit the soil type and the horses' grazing preferences, *plus* suitable fertilizer, may be applied. Germination rates seem to be better when semi-organic fertilizers are used, especially in a dry spring, as inorganic fertilizer salts can draw moisture *out* of the seed (by osmosis) just when the seed needs to draw water in!

Once the grass is established, it may be grazed (preferably by sheep) as described above (p. 139). Once the pasture is renewed it makes sense to pay attention to fertilizer policy and weed control, reseeding patches where necessary and harrowing and rolling as appropriate, and you shouldn't have to plough again for another twenty to thirty years, if ever!

Seed-bed conditions

A fine, firm and weed-free seed bed is the main requirement. The tilth needs to be fine and firm to facilitate an even sowing depth and enable the seeds, which are comparatively small, to come into close contact with soil particles. This is important as the small seeds contain relatively little stored food so rapid germination and early growth is essential.

The seeds are sown at a depth of 12–20 mm (approximately ½–¾ in) maximum, and not on the surface because of the dangers of dessication and bird damage, especially in a spring-sown crop. The seed bed must be weed-free (by ploughing or spraying) because many grasses are slow to establish and cannot compete against rapidly growing broad-leaved weeds or weed grasses for soil nutrients, water and especially sunlight.

Techniques

Direct reseeding

This is the traditional method in which the old crop or pasture is ploughed out (either in the autumn to sow next spring, or in late summer to sow in autumn of the same year) and the ground cultivated to produce a seed bed (e.g. by discing, rotavating, harrowing and rolling as necessary). It is most likely to be cost effective to the dairy farmer who sows in spring for an 'autumn bite' in the first year – unlikely to be applicable to horses who cut up young grass too badly with their hooves – but this method does tend to lead to reliable crop establishment. Late summer reseeding is to be preferred to get the crop established before winter and off to a good start next spring.

Undersowing

In this system, the herbage seed is sown at or just after (say three days) the time of sowing a cereal crop, for example a spring barley of the short-straw variety. The two are not drilled at the same time (differing seed sizes would lead to separation in the drill).

This method is perhaps more applicable to cheap (seed) short-term grass leys for conservation, although it is a reasonably popular method because it is cheap, in that the major costs of cultivation are carried by the cereal crop. However, there must inevitably be a compromise between the needs of the cereal crop and grass establishment. Usually a lower rate of seeding and fertilization is adopted for the cereal crop than when grown alone, with the result that the crop is below the potential of the land.

More important, perhaps, is the potential loss arising from poor establishment of the herbage crop, either through competition for water in a drought, or damage if the cereal crop lodges (flattens) or mechanical damage during the (more difficult) harvesting of the cereal crop. So, on the whole, this is not the method of choice for establishing permanent pasture for horses.

Direct drilling

Also known as 'sod seeding', in many situations this will be the method choice for establishing new pastures for horses.

Direct drilling may be used on stubble after arable crops or to renew a 'worn out' pasture, and involves minimal disturbance to the soil. Having said this, it may be desirable to 'Paraplow' if a pan has developed beneath the surface, for example after monoculture of cereals, or on grass leys after heavy stocking with cows (or horses) during wet weather, or after heavy silage making or arable machinery has caused compaction.

These pans often lead to waterlogging in winter or spring, and may prevent the grass roots growing down to where the moisture is in the summer months. Symptoms in an affected field down to grass include the spread of weed grasses such as watergrass, meadow grass and tall fescue. The field poaches badly in wet weather, and the sward may stop growing in the dry summer months and burn up completely in a drought. Broad-leaved perennial weeds such as docks and dandelions move in to the pasture. (This sounds like a typical horse paddock!) Compaction is of course not the only reason for such symptoms.

If, after expert guidance, you find your field is suffering from compaction, the Howard Paraplow provides a way of removing pans and subsurface compaction layers and improving surface drainage without disturbing the surface unnecessarily. It lifts and 'stretches' the soil down to a depth of 30 cm (14 in) breaking up any pans by cracking the soil along its natural planes of weakness. Increases in soil temperature of up to 4.5°C (8.1°F) have been reported following the use of the Paraplow during the previous September, giving an earlier start to growth of the pasture and, consequently, earlier access to spring grass.

When direct drilling, the Paraplow is used as follows. Spray off existing crop remains/weeds/herbage using a paraquat or glyphosate type herbicide (e.g. Gramoxone 100 or Round-up). Then Paraplow the field (starting the day after spraying). Allow the soil surface to settle for a few days, then either direct drill or surface cultivate the field by disc harrowing and then rolling, then direct drill the grass seed.

The advantage of using a combined seed and fertilizer drill is that of placing the fertilizer in the root zone, but with the disadvantage – especially when sowing in a dry spring – of the close proximity of the fertilizer tending to dry out the seeds, especially if inorganic fertilizers are used.

If Paraplowing is not necessary, the above system is the same for direct drilling, but normally a ten to fourteen day interval is allowed after spraying (with Grammoxone 100) before the seed is drilled. The soil should be reasonably dry and friable before drilling. The seed should have been carefully stored in a cool, dry place. Check and calibrate the drill for the appropriate seed rate and roll or harrow after drilling to close the slot if necessary.

It is a good idea to protect the new grass from slugs by mixing 7.5–11 kg/ha (7–10 lb/acre) of slug pellets (check specific manufacturers' instructions) with the seed in the drill, or alternatively, broadcast 15 kg/ ha (14 lb/acre) by hand or fertilizer distributor as soon as any damage is noticed. (Slug damage is indicated by grass blades broken off at soil level and thin shredding of leaves, especially at the base.)

Direct drilling is much quicker than other methods and is less labour and fuel intensive. On most equine establishments, the use of a contractor will be called for as with other methods.

Herbicides

Whether you plough or direct drill, it may be advisable to use a herbicide as the grass becomes established to kill broad-leaved weeds, as conditions which are good for the germination and establishment of grasses and clovers are also ideal for weeds – usually fast-growing annual weeds. The application of herbicide shortly after the emergence of the grass (and clover, if present) minimises weed competition, and allows the grass to become more quickly established and productive. Figure 15 illustrates growth stages for the application of herbicides in a sward.

The choice of herbicide will depend on (a) whether clover is present, (b) whether herbs are present (which is one reason why it may be more

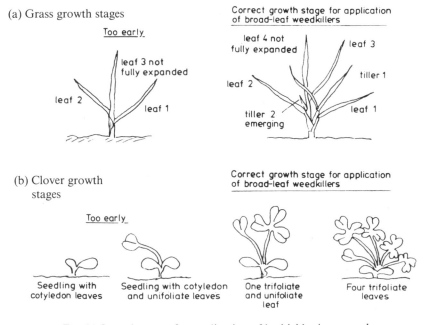

Fig. 15 Growth stages for application of herbicides in a sward.

convenient to plant a herb strip, or headland, at a later stage) and (c) the dominant weed problem. Most products have a very wide spectrum of varieties of weeds controlled, but they usually have one or two weeds which they are particularly effective at controlling at specific growth stages. Table 11 illustrates herbicides appropriate to various problems.

Table 11 Selection of herbicides

Problem	Herbicide
Weeds in pure grass	Usually based on hormones such as 2–4D, MCPA or MECOPROP, available on their own or in mixtures, and sometimes supplemented with other active ingredients.
Weeds in a grass and clover mixture	Usually a hormone based on 2–4DB, MCPB or low-dose MCPA, again available as straights or mixtures, and may be supplemented with other active ingredients.
If desirable herbs are present	Not much hope, but check with the chemical manufacturer. If weeds are patchy, you could try 'spot' spraying with a knapsack, or using a 'weed wiper' and missing the herbs!

I would not worry unduly about any dangers to stock from using a herbicide at this stage, as by the time the sward is established enough for grazing or silage making, all residues should be gone. *Do* leave an appropriate interval before 'topping' with sheep!

Sowing

There are various methods of sowing grass seed to produce a new pasture.

Broadcasting

Broadcasters may range from seed barrows, through fiddle drills to fertilizer spreaders, which must be recalibrated correctly for the small size of seed. They are most likely to be used if the land has been mouldboard ploughed in the traditional way; the seed must be harrowed and then rolled in after application. The seed rate will tend to be higher than for direct drilling.

Drilling

(a) Corn drills – these may have a fixed spacing of 15 cm (7 in), and as grass seed is generally drilled at half this distance, it will be necessary to cross drill at right angles, or remove the spouts and 'broadcast' the seed, rather like a seed barrow.
(b) Grass drill – specifically designed for the job and likely to be used by a specialist contractor or livestock farmer.
(c) Direct drill – a contractor's machine which is very accurate and facilitates the use of a lower seed rate.

Fertilizers

Depending on soil analysis, the average UK seed-bed application in spring is in the order of 75–125 units/ha (30–50 units/acre) nitrogen, 125 units/ha (50 units/acre) phosphate and 125 units/ha (50 units/acre) potash; for example 12.5 bags/ha (5 bags/acre) of a 10:10:10 NPK fertilizer. Less nitrogen would be applied in autumn to avoid wastage due to leaching, so for example a 10:20:20 would be used at this time, with a top dressing of nitrogen in spring.

Early management of a new sward

The first year is extremely important and can 'make or break' a new sward as far as being an easily managed and effective horse pasture is concerned.

For some considerable time, the soil will be more susceptible to treading and poaching, especially after ploughing, less so after direct drilling. If you have reseeded after grass there may be old rootstocks in the soil, or grass and weed seeds turned up by cultivation, and it is important that the young grass has the chance to compete against these.

The main aim is to encourage tillering of the grass, i.e. production of side shoots (tillers), which may in turn form shoots (tillers), to produce a densely tufted plant which will help to give the sward a good 'bottom'. Tillering may be stimulated by cutting, rolling or treading, particularly by sheep which will not tear up the immature plants, as horses or cattle might. For autumn-sown crops, the best management is probably to graze first (the following spring) with sheep if possible, or later with cattle combined with regular topping of the paddock. Regular balanced application of fertilizer throughout the first season will aid sward establishment and provide good keep for livestock.

I would avoid introducing horses until the late autumn of the next year for an autumn-sown crop, or the following spring for a spring-sown crop. Ideally, excitable young horses which tend to gallop and skid about and cut up the pasture should be avoided at first, both for the sake of the pasture and their own limbs.

Do not take a hay crop in the first year unless a conservation ley mixture has been sown specifically for the purpose. If desired, however, a silage cut at an earlier stage is acceptable. If you cut the grass or graze before flowering this induces tillering and, therefore, promotes yield and density of the sward. If you leave it for a hay cut, it grows upright and regrowth is from old leaves and stems, not tillers, so the sward doesn't thicken or develop a 'bottom'. This will also occur if the grass is not grazed or cut effectively during growth. Do not, however, attempt to take several silage cuts off a new paddock as plants will die off, especially if grazed or cut again within 21–24 days, because the root reserves are lost (in cutting) as the crop doesn't have time to rebuild reserves by photosynthesing.

A grazing or cutting programme should allow about 24 days rest between cuttings (14–15 days in a good spring) and on/off or strip grazing may be needed.

Pasture upgrading and renovation

With horse paddocks, it is usually more desirable to upgrade the existing soil/sward than to plough up and start again. *Most* horse paddocks are permanent pasture, although leys (grass grown as a crop for one, two, three, or four years, with the intention of ploughing up) may be grown specifically for conservation and used for grazing at the same time. The latter can be a useful source of 'worm-free' grazing for youngstock, followed then by older horses, although they can do a lot of damage to a young ley with galloping hooves.

Upgrading existing pasture

Henry Wynmalen points out in his excellent book *Horse Breeding and Stud Management* (first published in 1950) that 'moderate grassland can often be improved beyond recognition by good management and by good manurial treatment'. Also 'Good grassland can be maintained in that condition only by continuous good management, allied to adequate manurial treatment'.

A number of techniques can be practised, but the benefits of all will be short-lived if the drainage is not adequate and if the management is not changed to prevent the sward from deteriorating again.

To upgrade a pasture, (a) first it should be cut short, (b) it can then be allowed to regrow, and herbicide used on any new weed growth, or go straight to (c) harrowed, then (d) suitable fertilizer may be applied and (e) new grass seed of desirable grazing varieties may be broadcast, direct drilled or slot-seeded; then (f) rolled with a flat (i.e. non-ribbed) heavy roll, unless compaction is a problem. The new seed should then be given time to become established before grazing, or preferably topped again to promote tillering and *then* grazed.

After that, make sure you do not under- or overgraze or cut. Be vigilant about weeds and keep your fertilizer practice up to scratch. Harrow frequently and roll only if necessary if the soil is cloddy or puffy, or after poaching, or to press down stones before silage or hay making. Broadcast a few seeds on any bare patches which develop, and you should have a balanced and productive pasture.

On the reseeding side, perhaps I should explain the terms in (e) above.

(a) Broadcasting seed is using a broadcaster (or by hand) to scatter the seed onto the soil surface, and probably doing the same with the fertilizer. Inorganic fertilizers would have less of a check on germination using this method than direct drilling. Broadcasting seed is probably most effective on bare paddocks (after harrowing) with a surface tilth and some luck with the rainfall. After broadcasting the seed, harrow again, then roll very slowly with a Cambridge (ring- rib-) roller afterwards.

(b) Direct drilling can be tough in a living sward, and seedlings may have trouble fighting through the existing sward, if it is fairly dense, to become established. It is probably the best method for clay or wet soils. The fertilizer may be 'combined drilled' at the same time, and a semi-organic one would be more desirable due to its close proximity to the seed.

(c) Slot-seeding (fig. 16) is a relatively new technique using specialist equipment (see plate 19), whereby the existing grass is grazed as low as possible, then chemical (paraquat) is sprayed by the machine in narrow strips at predetermined intervals, e.g. every 40 cm (18 in) across the field. This kills strips of herbage which would compete with the new seedlings. If the sward is over five years old, insecticides against frit-fly and leatherjackets may also be applied. At the same intervals the machine cuts a small V-shaped slot, pops in a seed and the fertilizer (again semi-organic will give the best germination rates), and you follow up with a heavy roller

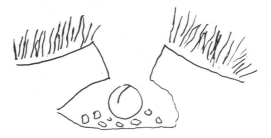

Day 1 : Seed lying in underground seedbed formed
         by action of slot seeder in lifting turf and
         shaking soil particles from roof of slit

Day 2 : Herbicide spray (e.g. Paraquat) causes
         destruction of pasture alongside mouth
         of slit. Reduces competition to
         germinating seed

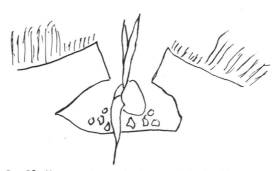

Day 20 : New seed germinates undisturbed to
         quickly rejuvenate pasture with new
         grass strains.

Fig. 16 Slot-seeding.

Plate 19 The slot-seeder provides a useful means of upgrading existing pasture. (*Photograph kindly supplied by MIL Ltd of Wolverhampton.*)

to close the slots. Stock must be kept off for at least three weeks to allow the new grass seedlings to become established.

You can use the last method to introduce, say, a high yielding ryegrass (or even herb strip!) and upgrade the value of the sward, although in clay soils the slot may become smeared and reduce germination, or heavy rain immediately afterwards can wash the seeds out of the slots if you aren't too diligent about the rolling. This method *can* be used without the herbicide, but results will not be as good.

A great deal of damage is done to pastures by putting horses on too early in the year, before the grass really gets going, so if you have a large enough acreage (hectareage?) it is a good idea to sow one paddock with 'early' varieties and keep it *just* for an 'early bite' and conservation.

Whichever method you use, there is *no point* spending money, time, effort and resources on upgrading a pasture if you do not intend to *keep* it upgraded! As one manufacturer of semi-organic fertilizer points out 'The fertilization of stud farms is clearly a subject of prodigious horizons, where stock of immense value, and other factors not normally associated with general agriculture must be taken into consideration', and this applies to all aspects of pasture management for equines.

For any grass-kept horse or pony, regardless of monetary or emotional value, the pasture is likely to provide up to 100% of its nutrition, up to 100% exercise, and be a major factor in its life. If you look after your pastures they will help look after your ponies, but don't expect something for nothing!

Conservation of forage for horses and ponies

Parts of this section are extracted from a series of articles by the author of this book, which appeared in *Equi* magazine during the period 1981 to 1983.

Silage, hayage and hay are three basic types of 'conserved forages'. All three techniques, involving pickling or drying, are aimed at preventing enzyme action and life processes both in (plant enzymes) and on (bacteria, moulds/fungi) the crop, and at retaining quality to as near the quality at cutting as possible, i.e. reducing loss in feed value.

The purpose of silage making is to encourage fermentation of the right sort, to 'pickle' a relatively wet crop (usually grass or maize) in its own juice. Hayage making also involves fermentation of a much drier crop, whereas the purpose of hay making is to stop fermentation of any kind in a drier crop. With silage the aim is to store it at around 20% dry matter (DM). Hayage is cut when the crop is more mature, and has a DM content of around 50% and hay is made from the mature crop cut

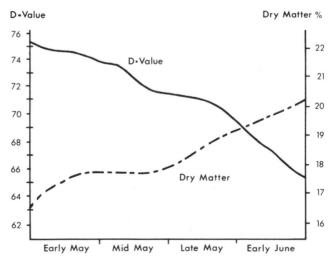

D-Value = Digestibility value based on DOMD% (Digestibility of Organic Matter in Dry Matter%).
For a later maturing grass variety the shape of the curves would be the same. D-Value is normally applied to ruminants but does give some guide to relative feeding value for horses.

Fig. 17 D-value plotted against yield.

preferably at around 20% DM (15–35%) and wilted and dried naturally or artificially for storage at 80-90% DM. Hayage is nearer to silage in its production technique, and between silage and hay in feeding value and digestibility. Hayage and hay are progressively less digestible than silage, as they are *usually* made from an older, more fibrous, less easy to digest crop.

The graph in fig. 17 shows a typical growth curve for a temperate grass, with the curve for digestibility superimposed. The object of conservation is to choose the optimum yield (tonnes per hectare) combination. The horse may obtain more nutrient from a given area of land by cutting a smaller crop at a higher digestibility level.

Hay, silage or hayage crops for which a mixture of grass varieties or legumes are used are likely to have different rates of maturity, and hence different fibre contents, which means they will ensile/cure at different rates, making them more difficult to conserve effectively. The 'heading' date gives a guide to the rate of maturity of particular grass varieties under average weather conditions.

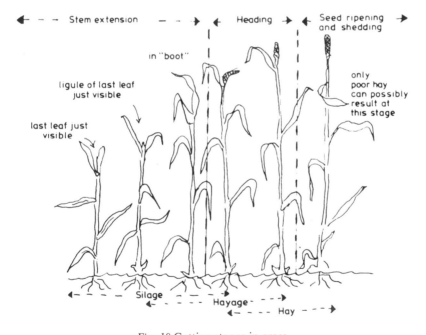

Fig. 18 Cutting stages in grass.

Figure 18 shows part of the Feekes scale of growth stages in grasses and cereals which gives a precise means of identifying different stages of growth in the crop. As a guide for cutting times in grass, silage is usually taken at or just after heading, hayage between heading and flowering and hay at flowering, although in practice most hay is cut later than this.

Traditionally, in the UK, we tend to think only of grass, grass/clover mixtures, lucerne and sainfoin (plus cereal straws) as forages for horses, although other possibilities include whole-crop cereals (cut when they are green), e.g. oats, barley, rye, wheat, maize, sorghum (milo), millet, etc., and cereal/legume mixtures such as oat and vetch. Fenugreek is another legume which I think we shall see more of in the future, especially as a specialist crop for horses. It is similar in appearance to lucerne. All these forages also lend themselves to the other significant form of conservation, which is drying to form meals and pellets (e.g. dried grass) – a so far underutilised feed source for horses.

Silage

Silage is any feedstuff which has been 'ensiled' i.e. placed in a silo (pit/clamp, sealed tower, or even large polythene bags) and allowed, or encouraged, to 'pickle in its own juice'. The process has been known since Roman times, although it has only relatively recently superseded hay as the major forage system in the UK. However, the acreage of land used for silage making has expanded rapidly in the last two decades as machinery, techniques and grass varieties have improved. In fact, grass and maize (corn) silage have been used for many years for horses in Europe, and many livestock farmers in the UK and USA are using it for their horses, but without the benefit of any precise published work to guide them.

The most promising method of silage feeding for horses is to use 'bagged' silage, i.e. the crop is baled usually in large (jumbo) round bales, and sealed in heavy gauge plastic bags, which may be stored adjacent to stabling or paddocks ready for use. They do need some machinery for handling. If this system is used, the bags should be tightly closed each time they are opened to prevent spoilage (plate 20).

When the cut crop is first ensiled, its life processes (metabolism) are still operating. Respiration and photosynthesis are the major processes concerning us here. As the crop dies its nutrient content is released to the micro-organisms which have accumulated on and around it – first the sap, then the more fibrous parts. Crop death in any conservation process needs to be as rapid as possible, as while the crop is still respiring, but no longer able to take up water and nutrients from the soil through its roots, the carbohydrates, important energy sources, are being used up in the respiration process faster than they are replaced as a result of photosynthesis, so the feed value of the cut crop is depreciating. Crop death is, therefore, hastened in the field by 'wilting' – regular turning to expose the whole crop to warm drying winds (if the weather is right!), and

Plate 20 Big-baled bagged silage must be tightly sealed to maintain feed value and is often safest when treated with a preservative. (*Photograph by Jill Forster*).

conditioning or laceration to aid release of nutrients to the micro-organisms and give them a larger surface area to work on.

The micro-organisms present come from the crop surface (splashed up from the soil, dropped there by grazing livestock and so on). They are harvested with the crop, which has been cut using a mower and is picked up and chopped by a forage harvester. Some micro-organisms are desirable (lactic acid producing bacteria) for the ensilage process, some are not (butyric acid producing bacteria, which give spoiled silage its 'rancid butter' smell). These butyric acid bacteria are 'obligate anaerobes' i.e. they can only be active and multiply in the absence of air. Most of the desirable bacteria are 'facultative anaerobes', i.e. they can thrive with air (aerobic conditions) or without it (anaerobic conditions).

When the crop is first cut and clamped (or bagged), air is present so the desirable lactic acid producing bacteria can thrive whilst the butyric acid producing bacteria are dormant. This gives the 'goodies' a head start. However, as they multiply and grow, they, and the residual crop respiration, are using up the oxygen, creating an anaerobic situation, which means the 'baddies' can start to become active.

Over the first few days the balance of desirable bugs present changes as the crop becomes more acidic due to fermentation of sugars (carbohydrates) and subsequent production of lactic acid by those bacteria. The pH is reduced to 4; the object of good silage making is to achieve this degree of acidity as rapidly as possible *without* encouraging the butyric acid bacteria. Most of the lactic acid is produced after day three, and if pH 4 can be reached by day six the crop is well-preserved. If the pH is not down to 4 by this time, the butyric acid bacteria start to grow and convert lactic acid to butyric acid, *and*, more importantly, break down the plant proteins, which are what we are trying to preserve! The proteins are broken down to amines such as putrescine, which is toxic to animals. The 'goodies' also break down proteins, by up to 50–65%. However, they are broken down to amino acids which may be reconstituted to microbial protein, which is also nutritionally desirable for the horse. This process, therefore, has no adverse effect on the crude protein content of the silage (which is really a measure of the nitrogen content). Unfortunately, the harmful amines in spoiled silage also contain nitrogen so look good on the crude protein part of a silage analysis.

Both lactic acid and butyric acid are 'volatile fatty acids' (VFAs) and are both present in the large intestine of the horse, produced by similar micro-organisms, as a normal result of the digestive process.

Legumes (lucerne, fenugreek, clover, vetches, etc.) contain larger amounts of organic acids than grass. These organic acids act as buffers (acid 'balancers') and make it harder for the micro-organisms to bring the pH down to preserve the crop, as they have to produce more lactic acid to overcome this buffering action. For this reason, legumes and legume mixtures are more difficult to ensile successfully than grass alone.

Moulds and fungi can also break down protein and spoil silage, but as they are obligate aerobes they must have air to do so, which is why exclusion of air is vital during and after silage making, by compaction, covering and sealing the crop. Compaction is by far the most important method of air exclusion in clamp silage. Sealing in towers or plastic bags is also of interest to horse owners, and it should be noted that if bags are torn or punctured during transport and storage, air will get in, resulting in rapid deterioration of the silage.

Another advantage of wilting and thereby drying the cut crop is that, if it is very wet, the lactic acid will be diluted and it will take longer for the lactic acid bacteria to bring the pH down to an acceptable level, giving the butyric acid bacteria a better chance to gain a toe-hold and spoil the silage. It is not usual to graze a silage crop before cutting as droppings would spoil fermentation.

Hayage

Hayage is the generic name given to forage crops, usually grass or grass/

clover mixtures, stored at around 50% DM (hay being nearer 80% and silage nearer 20%). Horsehage is the brandname of a form of hayage packaged in easy to handle units for the horse owner, the first of its kind, and other brands include Pro-Pack and Hygrass (which is produced by a somewhat different method). Horse owners wishing to produce their own hayage are advised to use a preservative in view of the difficulty of excluding air in an on-farm situation.

Hayage may also be stored 'loose' in air-tight tower silos, e.g. 'Haylage' in the Howard Harvestore, and although this is suitable for horses, the main difficulty is that you cannot readily sample the product to check its feeding value until you are actually extracting it to feed.

Hayage is cut later than silage, but before the crop has reached the hay-making stage. The yield of dry matter per hectare is consequently higher than for silage, and digestibility lower, because of the older, more fibrous material. However, it is considerably more digestible than most hay. In this process the object is again to exclude air. Hayage is really dry silage rather than wet hay, and the fermentation process involved is similar to ensilage. However, for tower silos, the crop is generally cut shorter to allow a greater release of plant sap for fermentation.

In hayage, fermentation down to the desired pH is more rapid than in wetter silage, as there is less moisture to dilute the lactic acid.

I have recently been asked if you can bale wet hay to make hayage and store it in bags, but I would be very dubious about this as it is likely to be of poor feed value and spoil easily. If you have a wetted hay crop, however, it might be worth a try, rather than losing the crop, *provided* you apply a preservative as you bale it. Check quality *very* carefully before feeding. It is far better to cut the hay crop and barn dry it if you can.

Hay

For those who continue to make hay for horses, there can be little doubt, that for the most part, a far better product may be obtained by use of artificial barn-drying techniques, which reduce weather dependence and enable the crop to be cut at the optimum time for the right combination of yield and feeding value.

Many horse owners have the mistaken idea that two year old hay is of superior quality for horse feeding, when in fact its vitamin content is often negligible due to inevitable deterioration even under the best storage conditions, and it is more than likely to harbour mould spores and dust which can set up allergies.

The object of haymaking is to reduce the moisture content of the green crop to a level low enough to inhibit the ultimately destructive plant and microbial enzymes from fermenting the crop. Coincidentally, this concentrates the nutrients so that fewer kilogrammes or pounds (fresh weight) need be eaten than fresh herbage to supply an equivalent amount of nutrients. To do this, the moisture content of the green crop, 65–85%, must be reduced to 15–20% to make hay. Unfortunately, in attempting to achieve this aim, the majority of haymakers tend to let the growing crop become overmature (when it is drier, stemmier and more fibrous) and consequently less nutritious, and less digestible.

Horse hays used in the UK and Ireland generally contain 4–10% crude protein. Oat- or barley-straw is more nutritious than the lower quality hays! Protein digestibility tends to be lowest in the hays of lower crude protein content, rendering them doubly unsuitable, and yet many horse owners who would be furious at the inclusion of so-called 'fillers' in a concentrate ration will adamantly feed poor hay rather than resort to straw.

Hay of mixed varieties (e.g. meadow hay) is rarely at the right stage of maturity all at once, leading to very variable quality and uneven drying which can lead to storage problems. Weeds are also more likely to be a problem, including highly poisonous ones such as ragwort, avoided by grazing animals but often eaten dried in hay, with usually fatal results – a surprisingly common form of severe illness and death in horses.

The main factors which determine the feeding value of conserved forages are the type of herbage from which the hay is made, soil type and fertilizer treatment, pest, disease and weed control (if necessary) in the growing crop, the stage of growth when cut, weather conditions at cutting and throughout the growing season, the actual method of harvesting and drying, and subsequent storage conditions. In the UK, the D-value is a measure frequently applied to forage, and refers to the amount of digestible organic matter in the dry matter. For hay, a D-value of 68 is very good, 63 acceptable, 55–60 average and less than that all too common!

The Ministry of Agriculture, Fisheries and Food (MAFF) in their Advisory Bulletin no. 448 defines grass hay quality as follows:

First quality – young, leafy grass cut when very few flowering heads have appeared; D-value 65–70%.

Second quality – headed grasses in early flowering stage but with a good proportion of leaf; D-value 60–65%.

Third quality – grasses cut in full head *at the traditional hay stage*; D-value 55–60%.

Fourth quality – mature grasses that have shed some of their seeds leaving a rather fibrous stem; threshed grasses grown for seed production fall at the lower end of this category; D-value less than 55%.

It is worth remarking that, on a dry matter basis, the feeding value of hay can never exceed that of the crop from which it is made because there are always losses incurred during the making. A frequent misconception is to overestimate feeding value just because hay is well made, and farmers and horse owners alike are surprisingly bad at guessing the feed value of hay.

Losses during harvesting occur largely to those parts of the crop which are highest in feeding value (generally leaf material lost by shattering) and the extent of these losses depends on the method used for making the hay and the length of time taken. They are smallest when the crop is properly barn dried, but they may also be small when the hay is made quickly in very favourable (and unusual!) weather by a ground-curing technique based on early conditioning.

Greater physical losses, and reduction in quality of remaining material, occur during 'normal' field curing, increasing further when the crop is exposed to rain, particularly when the rain falls during the latter stages of curing. Baling hay which is too damp will almost certainly lead to mould formation and dustiness, and even overheating and spontaneous combustion in the stacks – a common cause of farmyard fires. However, when a lesser degree of heating occurs, it leads to a reduction in palatability, and harmful effects of dust and mould spores on horses and man.

The Ministry of Agriculture, Fisheries and Food Advisory Bulletin no. 448 divides hay into the following categories for the purposes of rationing and I quote their comments, for dairy cattle, where applicable to horses:

'*Very good hay* – very little hay comes into this category. It has to be made from first quality material by a method such as barn-drying which involves small losses.' Traditionally 'ground-cured hay is very rarely of this quality. Very good hay will be eaten in large amounts if offered and may make a significant contribution to the animal's production requirements in addition to those of maintenance'. It may form the sole feed requirement of, say, a dressage horse, which needs to be in good condition, fit, but not above itself, when fed along with a free access supplement. 'Care must be taken that consumption of this valuable material is not excessive, and the ration is properly balanced to utilise any surplus protein.'

'*Good hay* – good hay is made from first quality material by ground-curing or from second quality material by methods involving very small losses,' e.g. barn drying. 'Hay of this type, when fed in sufficient quantities will cover the maintenance requirements and may provide a small share of the production ration.' This is the best hay most of us are likely to see, and most years are worth paying well for. Below this quality, in 'expensive hay years', alternatives should be considered.

'*Medium hay* – most of the hay made by traditional methods falls into this category' – particularly meadow hay, so often priced at a premium as horse hay. It may be better to pay a little more for hay and balance it with oat straw if the protein is too high. 'It is made from third-quality material by ground-curing methods which do not involve serious losses, or from first or second quality material which has been poorly made. This type of hay when fed to appetite will provide a … maintenance ration for most classes of stock.'

'*Poor hay* – poor hay is either well made from fourth quality material, or badly made from higher quality material. The unsatisfactory harvesting can be the result of excessive exposure to rain, to over-heating, or to moulding after carting' (removal from field and stacking). 'It is important to remember that overheated hay, although (it can be) palatable and attractive to stock, has a low digestibility and is always poor. When fed in normal amounts poor hay does not supply enough nutrients even for maintenance purposes … Poor hay is often lacking in minerals such as calcium and phosphorus and balanced minerals should be fed; an 'additional' Vitamin A and D supplement should also be included if the hay has been badly weathered.' Where poor hay is the only forage available, this is where I feel cod liver oil is a useful supplement over and above any normal supplement usage, and this is one of the few instances where I would use cod liver oil by choice as a supplement. It is often regarded as a simple, therefore general supplement, whereas I see it as a concentrated, specialist and highly specific supplement, and insufficient for general supplementation purposes.

Hay of any sort which has become mouldy should *never* be fed (or even used as bedding). Mowburnt hay will have had most of its vitamins destroyed, and hay which has been rained on, even only once, has soluble nutrients (vitamins, minerals, trace elements and sugars) washed out of it.

In recent years, chemical preservatives have been increasingly used on hay. Some scaremongers predict dire consequences of feeding treated hay to horses – a rather blinkered attitude. There are several different brands of preservative on the market, mostly based on proprionic acid, a VFA, which is produced naturally during digestion in the large intestine of the horse. Trials undertaken in the USA report better results using proprionic acid-treated alfalfa (lucerne) hay than equivalent untreated hay. In the UK there is an acetic acid (cider vinegar) plus molasses preservative which is likely to be highly palatable to horses.

Certainly preservatives should be used on any form of bagged silage or hayage for horses to preclude the possibility of botulism.

The object of the preservatives is to retard or prevent formation of moulds on hay, so they are particularly useful in a wet or catchy haymaking season. They may be sprayed on the cut grass before or during baling. However, proprionic acid does destroy vitamin E so a vitamin E containing supplement should be given. This vitamin destruction is sometimes given as a reason for not feeding proprionic acid treated hay (or cereals) but when the alternative is mouldy, poor quality hay, which is probably low in vitamins and soluble nutrients anyway, plain logic and horse sense should make most people favour the preserved product plus a supplement every time. Few people are likely to apply a preservative to hay that doesn't need it, as this would be a needless expense. People who are concerned about a cumulative depletion of vitamin E reserves when preservative treated hay is fed over several seasons are, to my mind, placing a rather over-confident reliance on the vitamin content of 'normal' hay and bodystores of vitamins, and unless they are consistently exceptionally lucky in their hay supply, or unusually good at producing hay, they should probably be looking at some form of supplement for any high performance horse, particularly of vitamin E.

As with all feedstuffs, high quality hay which costs more per tonne is usually better value for money than a cheaper, poorer quality product.

There is really no need to wait until hay is a year old before feeding it. Straight off the field there may still be chemical reactions going on and it needs time to settle down, but after that it may be fed with confidence, especially if bales are opened twenty-four hours in advance and spread a little to finally 'crisp' in the air. As with all new feedstuffs, new hay should be introduced gradually (or even a new batch of 'old' hay) and mixed with the existing supply for the first few days until the changeover is complete.

A problem with mixed grass/cereal and legume hays is that the legumes are usually more succulent than the grasses, so to 'make' the hay either the grass is overdry (leading to losses due to leaf shatter and dustiness) or the legume is damp, leading to mouldiness which can affect the whole sample.

Chemical changes and losses during drying

Apart from the obvious effects of weather damage mentioned above, the quality of a hay crop is subject to a number of influences.

Action of plant enzymes

Once the crop has been cut it does not die immediately, and natural chemical reactions such as photosynthesis may continue for some time. Ideal haymaking weather is dry and warm with a gentle drying wind. If the cut grass is turned frequently to expose it as much as possible, rapid drying will cause enzyme actions to cease. The losses incurred due to enzyme activity (which is desirable in the *growing* crop) are mainly due to respiration, and thus oxidisation of sugars (to carbon dioxide and water), i.e. the soluble carbohydrate fractions, and a consequent loss of digestible energy value of the crop. As a result of this, the cell wall constituents, cellulose and lignin, become more concentrated, which is why the crude fibre percentage in the DM is higher in hay than the uncut crop.

Proteins are also altered by the action of the plant enzymes. After cutting, proteins are broken down (proteolysis) by enzymes to their constituent amino acids. Provided the crop is not exposed to rain or heavy dew, the crude protein value will be maintained, but if it becomes wet, soluble nitrogenous compounds may be washed out. Nutritionists Macdonalds, Edwards and Greenhalgh point out that 'the actual crude protein content of hay may be by itself a poor guide to the nutritional value, since it may remain the same, or even increase (relatively speaking) as a result of losses of other nutrients'.

Oxidation

The visual effect of oxidation in field dried hay is the bleaching due to destruction of pigments. This is very important as many equestrian writers claim hay to be a good source of beta-carotene, the precursor of vitamin A. However beta-carotene is also a (red) pigment, and it may be reduced from 150–200 mg/kg (ppm) in the DM of fresh herbage to 2–20 mg/kg (ppm) in the hay. Rapid drying helps to prevent this loss, and although I have seen claims in the equestrian press that barn drying 'destroys vitamin A content', in fact it helps preserve it, or at least the beta carotene, and losses can be reduced to as little as 18%. There is a certain ill-informed antagonism to anything 'new', common in equestrian circles, in these claims.

Where barn drying does lose out compared with field drying is in vitamin D content, as sunlight irradiates ergosterol in the green leaves and changes it to vitamin D. However, this is generally not a sufficient reason against barn drying, unless you can guarantee perfect haymaking weather, and long-term storage (popular for horse hays) means the vitamin D in the fresh hay has substantially deteriorated by the time the hay is fed.

Leaching

The effects of leaching by rain of soluble nutrients are most pronounced when the crop has been partly dried. Rain immediately after cutting can be less damaging in this context than after several hours of good drying weather. Losses are of soluble carbohydrates, nitrogenous compounds, vitamins, minerals and trace elements, and by concentration lead to a higher crude fibre level in the dry matter of the hay. The moisture can also prolong enzyme activity in the cells and provide an environment which encourages mould growth.

Mechanical damage

Leaves, the most nutritious part, dry more rapidly than the stems, but the stems must be dried properly if the hay is to store well. Because of this the leaves tend to become overdry and brittle, and will shatter if roughly handled. Rough handling includes mechanical techniques (such as crimping) which are being applied to hasten the drying of the stems, and even the action of turning or windrowing the swaths, and particularly baling. (Figure 25 shows the rather violent route hay travels when being baled.) If loss of leaf is excessive a considerable reduction in feed value may result. Modern haymaking machinery is designed to reduce this problem, but unfortunately a lot of non-farming horse owners rely on archaic second-hand machinery thinking it is cheaper to make their own hay this way. Often it would be more cost effective either to buy better machinery and make better hay, or pay a good contractor equipped to do the job properly. Much makeshift, make do and mend machinery is perfectly acceptable for practical purposes, but worn out hay (and silage) conditioners are almost always a false economy.

Micro-organisms

Bacteria and fungi on the growing crop, carried by wind or rain, splashed up by rain from the soil and so on, may never become active if a crop is dried rapidly. If, on the other hand, drying is prolonged due to adverse weather conditions, these bacteria and fungi (including moulds) are likely to become active. Mouldy hay can be harmful to the horse and man (at worst causing 'Farmer's Lung', a debilitating disease caused by *actinomycetes* on the hay), and is at best unpalatable and of poor feeding value. Preservatives, such as proprionic acid, have a fungicidal action and prevent mould formation.

Changes in the stack

Chemical changes and losses frequently continue once the baled hay has been stacked, especially if it is not properly dried. Stacked hay may contain anything from 10–30% moisture and at higher levels micro-organisms, and to a lesser degree plant enzymes, will continue their activities. Apart from damage and alteration of sugars and proteins, losses of beta-carotene may be considerable if temperatures within the stack exceed 5°C (41°F). Overheated hay (blackened or brown) has very poorly digestible crude protein, as little as 2.6%, so the protein contribution of this diet would be negligible – in fact newspaper and corrugated cardboard are more nutritious! If hay is stored at less than 15% moisture, little or no deterioration will occur.

If you intensively fertilize a field of grass, of varieties selected for intensive production, weather permitting you should get at least two or three cuts of highly nutritious hay, provided of course you do not seek to graze it as well. A mixed grazing/conservation paddock normally will only be expected to produce one reasonable hay crop.

All the fertilizer companies produce useful booklets about grassland management, and I have found the UKF 'Swardsman' system especially well presented, adaptable and easy to follow. As an example, given a good basic starting analysis of the soil (i.e. with lime (pH) and basic slag treatments up-to-date, and a good nitrogen index) and an appropriate seed mixture, they recommend $2\frac{1}{2}$ bags of Swardsman (NPK 25:5:5) per acre before the first cut, giving 63 units nitrogen, 13 units phosphate and 13 units potash per acre; then $2\frac{1}{2}$ bags UKF no. 10 (NPK 20:5:15) after the first and subsequent cuts (i.e. 50 units nitrogen, 13 units phosphate and 45 units potash after each cut, per acre). If you are going to graze or take silage as well or if the land is in poor condition, the system is adjusted accordingly.

Yields vary enormously and depend on soil type, fertilization, weather, time of cutting, moisture content, grass/clover varieties chosen, etc.; but basically the later you cut, the higher the yield in tonnes and the *less* digestible and, therefore, nutritious the hay, as the crop becomes more stemmy. However, you should expect around 4500–5500 kg per hectare (2 tonnes per acre) per cut under a reasonably intensive system.

Machinery

The field management equipment which is chosen for horse pastures will obviously depend largely on economics and the size of the enterprise. It may range from a scythe and a fork or shovel for the one-horse owner (and possibly a knapsack sprayer), through to harrows (which may be pulled by car, Land Rover or 'sit on' garden mower, or even the horse), mini tractors and 'motorbikes', and full agricultural grassland management tackle, which at the least would include a tractor, harrows, a mower/topper, fertilizer spreader and agrochemical sprayer. On the largest units, seed drills, hedging and ditching attachments, post drivers (for fencing) and hay or silage making equipment may be included. Ancillary equipment might include a customised vacuum for collecting droppings, and drainage or irrigation equipment.

Whether your equipment is a simple scythe or a tractor and all the tackle, it makes good sense to spend a little time and effort looking after it, both to ensure efficiency, safety and accuracy of operation, and a long working life. Because many horse people do not like machinery, they have a habit of neglecting it, which is unfortunate and can be both dangerous and expensive.

Wherever possible, keep equipment under cover and, of course, out of reach of livestock. Harrows left on a verge in spring soon become overgrown with weeds and forgotten about until your valuable yearling breaks loose and gallops over them.

If a piece of equipment has a handbook, *read it* before using the equipment, even if you have used something similar before. Keep it either with the machine, or in a suitable accessible container in the machinery/tool shed, or office, and make sure all personnel using that piece of machinery or tool know where to find it.

Follow the manufacturer's instructions with regard to oiling and greasing, both through the season and when storing overwinter and starting again in spring. Keep scythe and mower blades sharp, screws as tight as is appropriate, blow out sprayer nozzles (*never* poke them out with wire as this will widen the hole and upset the calibration) and brush or wash off excess dirt or grass after use. *Always* check calibration of sprayers, drills and fertilizer spreaders, and that no spouts or nozzles are blocked. Remember that if you change from an inorganic prilled fertilizer to a pelleted semi-organic one, you will have to recalibrate the spreader.

Tractors should be topped up with diesel at night (so that condensation doesn't build up in the tank as they cool down), and battery and radiator water and engine oil kept topped up but never overfilled. Keep tyres at the right operating pressure, and if they are filled with ballast (water) for any reason, do not forget to add antifreeze in winter, or remove it if not required.

Chain the two lower arms of a three-point linkage on the tractor together to prevent them crashing about, and always attach them in the following order: left, right, top, and detach them in the reverse order: top, right, left.

Check safety of operation of machinery – never operate near playing children or animals and do not allow passengers to ride on the towbar or side of a tractor. If you have passengers on a trailer they should keep their feet up or over the back, and never sit at the front or side with legs dangling as, in the former case, they may be jolted off under the wheels and in the latter, they may be caught by gate posts, etc.

Always keep tractors in low gear when going down hill, and if possible fit a cab or, at the very least, a roll bar. In many countries there is a statutory requirement for this, with certain exemptions for tractors which can be proven to be used around buildings with low overhead obstacles.

Check the height of fore-end loaders and bale-pick-ups going under power lines and do not let people ride on a fore-end loader. Ensure that weights are fitted, if required, to the front of the tractor – most likely when ploughing.

Obviously children should not be allowed to operate machinery, or indeed tools such as scythes, forks and wire cutters.

Machinery for large acreages

Assuming machinery is being obtained purely for use on a horse enterprise, I would say that a Ford 4000 (or equivalent) tractor with PTO

Plate 21 Tractors are one of the few items of machinery which can be purchased second-hand with reasonable confidence. Note the sturdy roll bar which is essential if a cab is not fitted. Attention to regular maintenance wil pay dividends in the longevity and usefulness of such a machine. (*By kind permission of Mrs B. Ehlers.*)

drive facilities (power take off for operating driven implements, e.g. mower/sprayer/spreader/rotavator, rather than just trailed implements, e.g. harrow/roller) would be adequate. A good second-hand tractor would probably be quite acceptable (plate 21).

For areas up to 20–30 ha (50–125 acres), one of the new four-wheel-drive 'all terrain vehicles', based on a motorbike, would be perfectly adequate. In fact these machines are most useful on any enterprise for field maintenance and even for 'walking' the course at events as they can be easily transported in a horse box (see below)! In addition, some sort of harrow would be essential, preferably a reversible spiked chain harrow with longer spikes on one side.

It is not usually a good idea to buy second hand fertilizer spreaders or especially agrochemical sprayers, as they can be worn and consequently virtually impossible to calibrate accurately. Always have some spare nozzles available for a sprayer, and mix and agitate the chemical as per instructions. Be especially careful where you pump out and dispose of

Plate 22 Herbicides should be selected, stored and used with care, and empty containers disposed of safely afterwards.

surplus mixed spray, and spray cans or containers. Never re-use agrochemical containers. Ensure that you follow the manufacturer's instructions with regard to wearing gloves, masks and protective clothing and wash your hands after use, whether or not you have worn gloves. You cannot be too careful with agrochemicals and a slapdash attitude is dangerous and stupid (plate 22).

There is considerable disagreement about the use of rollers in pasture. Some authorities say that in most cases they can do more harm than good, others that they are an essential aid to pasture management. In general, at present, I prefer in most situations to use a 'Cambridge' or ring- or rib-roller, which is lighter, but others would advocate a heavy flat-roller filled with water or even cement (plate 23). I feel the latter really is too heavy for most situations, but is an essential part of the establishment of new pastures. A common cause of poor seed establishment is inadequate rolling. It is probably useful to have a ring-roller available to break down poached areas after winter, or for preparation of new pastures, or if soil becomes 'puffy' after a frosty winter, but overuse should be avoided as it can cause compaction and damage to the soil structure. Whatever type of roller you choose, rolling is useless or even harmful unless it is done slowly, and roaring around a field with a roller in tow is simply a waste of time and fuel.

Plate 23 A heavy flat-roller (L) and a ring-roller (R), also known as a 'Cambridge'-roller or a rib-roller. Care must be taken to ensure that use of a roller does not cause compaction of the soil structure.

As far as choosing a mower is concerned, much depends on whether you require it just for 'topping' the paddocks to tidy up coarse growth or intend to conserve your own forage as silage or hay, and also on the acreage involved. Different types of mower are discussed below.

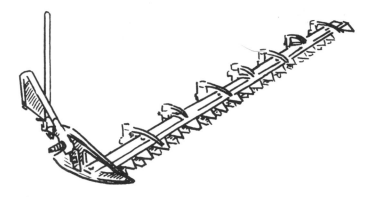

Fig. 19 Cutter-bar mower.

(a) *Cutter-bar mower* (reciprocating knife mower), developed from one of the earliest examples of mechanised farm implements, still remains the most widely used machine for grass cutting, despite the development of the flail and horizontal rotary mower (fig. 19). Basically, this mower consists of a cutter bar carried on a steel frame. The knife drive is powered by the tractor power take off through the medium of a counter shaft, or by a hydraulic motor.

There are various types and positions of attachment. These are:

(i) Semi-mounted type, which is supported partly by the draw bar of the tractor and partly by one or two castor wheels at the rear. This form of attachment improves manoeuvrability when compared with trailed machinery and avoids restriction and strain on the power-drive shaft universal joints.

(ii) Fully-mounted type, (rear attached) which is totally suspended on the tractor's three-point linkage so the whole mower has to be lifted for cornering or transport.

(iii) Fully-mounted type, (side attached), which is sometimes referred to as the 'mid-mounted' type, is positioned mid-way between the front and rear wheels of the tractor.

The big advantage of this system is the good visibility of the cutting operation for the tractor driver. In addition, for hay making, a crimper or tedder can be attached at the rear of the tractor and the two operations carried out simultaneously. (A crimper speeds drying of the cut crop, a tedder turns the rows for drying or can row up two rows into one – at night to reduce wetting by dew and rain.) On any type of cutter-bar mower, basic maintenance is important. A cutter bar that is out of alignment may result in poor cutting, heavy draught or breakages.

(b) *Flail mowers* are increasingly popular for silage cutting, and are less likely to be found on the horse farm than other types, though they can be can be useful if topping a long crop of grass if you do not wish to let it drop back and rot on the sward. If the pick-up flail mower is used, the cut grass can go straight into a trailer for removal.

The difference between a flail mower for hay and a flail harvester specifically for silage is mainly that the latter thoroughly lacerates the cut crop during the harvesting process. The term 'flail mower' is normally applied to mowers employing a horizontally arranged shaft with swinging cutting blades rotating at high speed in a direction opposite to that of the tractor wheels. Knives may be mounted on the cylinder in a variety of ways, the most common being a simple pivot mounting which allows the blades to fold inwards on meeting an obstruction or overload. The cut crop is thrown forwards and upwards against a steel hood, which is designed at the rear to reduce the velocity of the crop and drop it lightly in a uniform swath behind the rotor.

(c) *Rotary grass cutters* are probably, along with cutter-bar mowers, the most common on horse enterprises. They can cope with rough ground and tough, long grass rather better than the flail type mower.

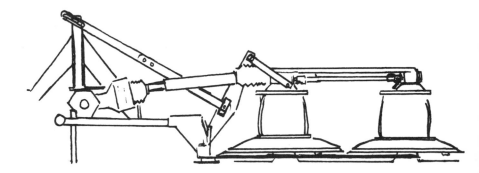

Fig. 20 Disc mower.

Rotary mowers are of two main types : drum mowers driven from the top by a shaft and gears or V-belts and disc mowers (fig. 20) which have the cutting mechanism driven by spur gears housed in a casing below the knives. Drum mowers tend to be larger than disc mowers. The rotors of drum and disc mowers are arranged in contra-rotating pairs. Power requirement is high compared with that of cutter-bar mowers and the purchase price is considerably higher. However, this may well be justified if you plan to conserve your own forage as well as using the mower for topping the paddock.

Disc mowers are basically lighter than the drum variety. Some machines, in fact, consist of two small outer and inner drums with discs between them combined. Many disc and drum mowers can be provided

Plate 24 A disc-mower suitable for hay-making on smaller acreages, can also be used for topping the paddocks when necessary.

with built-in attachments for crop conditioning (to speed drying) and again may be worthy of consideration if you are going to make your own hay (plate 24). However, the conditioned crop will spoil faster if rained on after treatment. Again basic maintenance and cleaning is advisable.

With regard to fertilizer distribution, a mounted hopper-type distributor, which broadcasts the fertilizer by spinner or an oscillating spout, should be perfectly adequate. The size purchased will obviously depend on the acreage concerned and number of times you expect to apply fertilizer in a year. The object is to achieve a uniform broadcasting of the fertilizer over the surface of the soil, with no gaps or overlapping, which will depend on the machine and the accuracy with which it is driven (plate 25).

Some fertilizer companies offer a spreader calibration service, which may be free to larger acreages, which provides a check on distribution pattern and a diagnosis of any reasons for inefficiency. When evaluating the cost (if any) of such a check, it should be borne in mind that there can be considerable loss of grass yield before visible striping of the sward occurs. Even new machines can be faulty due to a minor accident to a vulnerable part such as a delivery disc or a deflector plate.

With regard to spreading muck or FYM, in view of the unsuitability of horse manure (unless *extremely* well rotted) as a fertilizer for horse paddocks, due both to the potential parasitic worm problem and the unbalanced nature of horse manure for horse paddocks, it is unlikely that

Plate 25 A simple fertilizer broadcaster is a useful aid to balanced pasture management on larger enterprises. Care must be taken to ensure it is correctly calibrated.

a horse farm will require a muck spreader. Better to swap the horse manure for cattle manure from a local farmer who will spread it for you, or sell it to a mushroom grower. For horse pastures on a mixed farm, it is presumed there will be expertise and equipment on the farm side of the enterprise for spreading muck.

With regard to spraying herbicides, in the absence of trained assistance it is probably as well to use a contractor. It is certainly *not* a good idea to buy a second hand sprayer. A small, tractor-mounted sprayer will probably be quite adequate, and must be maintained and used strictly according to the instructions of the manufacturers of both the machine and the agrochemical in question. A sprayer which could well be of interest on all but the largest horse enterprise is the specially designed mounted machine for the Honda 4WD all terrain vehicle described below.

An alternative for controlling weeds which are taller than the grass is a weed wiper (described below) and tractor-mounted versions are available for larger acreages. They operate by literally wiping herbicide onto the weeds with an arrangement of rollers or wicks and a pressure regulated herbicide supply system. They are less versatile than a sprayer as they cannot be used to deal with weeds within the sward, nor can they be used for applying pesticides to deal with wire worm or leatherjackets. However, they can be used in windy weather, unlike a sprayer, and as far as safety is concerned are probably more 'idiot-proof' from an operator's point of view. Certainly the small hand-held weed wipers described below can be very useful for spot treatment on paths and around buildings. The tractor-mounted system is really better suited to root crops than grassland. However, if this type of machine is to be used, it should be remembered that front mounting can result in transfer of chemical from the weeds on to the grass by the tractor wheels and should not, therefore, be attempted.

Another item which will be most useful is of course a trailer, and the most versatile is one which can easily be converted from one with sides to a flat-bed trailer.

The other item of tractor-mounted equipment which might be considered is a vacuum cleaner especially designed to collect horse droppings from a paddock. This is only likely to be worthwhile as a means of worm control if you have the staff and equipment available to collect the droppings not less than every two to three days, and will never eliminate the need for control of parasites with anthelmintics. If you do decide to collect the droppings, it is certainly quicker and less laborious than using a trailer or wheelbarrow and staff with forks, spades or boards. Harrowing droppings will only dessicate them and kill worms in drying weather, so vacuuming droppings is certainly particularly beneficial in wet seasons or areas, when harrowing would simply spread viable worm eggs or larvae over a wider area.

The pros and cons of different hedgers, ditchers, irrigation equipment, cultivation and seed-bed preparation equipment and forage conservation and conditioning equipment are beyond the scope of this book.

Machinery for smaller acreages

For 20 ha (50 acres) or less, it is more difficult to justify the purchase of a 'full-size' tractor, and this is where the mini-tractors and ATVs (all terrain vehicles) come in to their own. These are a fairly recent innovation, and can be extremely useful both for field work and around the buildings, i.e. mucking out, moving jumps, harrowing and some can even be used in the garden to mow the lawn, etc. In fact the ATVs can be used as 'fun' runabouts around the farm and even for 'walking the course' at horse trials, etc., as they do not damage the soil structure (although I would foresee some limitations being put on their use for this purpose by some event organisers and landowners if they become *too* popular!).

There are many brands of mini-tractor, most of them Japanese, although a Massey Ferguson range of machines is also available in the UK. Most come with a range of customised tackle including rollers, harrows, fertilizer spreaders and mowers or toppers, and possibly a small trailer, and many also have cultivation equipment, diggers/ditchers, fore-end loaders and so forth.

Examples are the Massey Ferguson 1000 range of compact tractors, which are available in 16, 21 or 27 horse power models (MF1010, 1020 and 1030 respectively), with a choice of two- or four- wheel drive, the latter being particularly useful if you have to cope with difficult or hilly terrain. They are available with 'turf' tyres suitable for lawns and playing fields and agricultural ones with a deeper tread. There is also the Kubota, in a 17, 19 or 27.5 horse power tractor, diesel operated, which can also be obtained in two- or four-wheel drive. Both the Massey Ferguson machines and the Kubotas have three-point linkage and power take off facilities, and can be fitted with a safety cab or roll bar. Lely Iseki is another well-known range of mini (plate 26).

The advantage of the minis is that they are highly manoeuvrable and are especially useful for awkward areas. They may be employed as the

Plate 26 A Leli Iseki mini-tractor is useful on smaller acreages and around the yard.

sole source of power on smaller acreages, or as supplementary 'multi-purpose' vehicles on larger farms and estates. In general, the four-wheel drive versions operate with about 50% more traction, which can be useful in heavy ground, on slopes, and to pull heavier loads.

A disadvantage of the mini is that they are proportionately as heavy as a standard size tractor, a problem which is not shared by the rather unusual Honda ATVs. These have evolved from the Honda motorbike so are rather maxi-motorbikes than mini-tractors!

The big advantage of these ATVs is that they can be driven over wet or muddy ground without doing damage to the soil structure. They are true 'low ground pressure' vehicles (in fact I've tried this claim by letting one drive over my foot, wearing thin court shoes, with no ill effects whatsoever!). Apparently they can be readily driven pulling a full load on top of a field so muddy you or I would sink in to our ankles wearing wellingtons. They were initially developed as a three wheeler fun-vehicle, based on a motorcycle, but have gradually evolved into three- and four-wheeled versions, the latter as a two- or four-wheel drive machine, with a comprehensive set of customised implements.

The Honda ATVs do not have power take off or three-point linkage like a tractor, so equipment is either fully mounted (e.g. their custom-made sprayer designed to distribute the weight of the sprayer and water over both rear wheels) (see plate 27) or trailed by a drawbar and self-powered via the ATV's electrical system.

The ATVs are petrol powered with electric ignition, and with a bit of practice are easy to drive and manoeuvre, rather like a motorbike that cannot fall over! Steering with handlebars seems a little odd at first but the vehicle is comfortable, even over rough ground, and very nippy. There are racks at the back and front which can be used to transport bales or feed bags, or in the case of the rear one, as a mounting for the sprayer.

The engines are all single cylinder four-strokes and are easy to start, even in wet conditions. The air intake is just below handlebar level, which allows the machines to be ridden through quite deep water. Transmission is via a four- or five-speed motorcycle type gear box operated by the foot. Some models have a dual ratio, the low range being extremely useful for towing heavy trailers or broadcasters.

Other features include (on the ATV 250 'Big Red') a detachable headlight which can be used as a work/search light, and built-in storage containers. The TRX 250 is a four-wheeled version of the 'Big Red'. Accessories include a trailer chassis fitted with LGP (low ground pressure) wheels and a 5 horse power Honda stationary engine which is set to run at 540 rpm (as would a standard tractor PTO shaft). This chassis is designed to take both a sprayer and fertilizer broadcaster, which are

Plate 27 The Honda ATV, shown here with a customised mounted sprayer fitted, is based on a motorbike. Available in three- and four-wheeled models, this versatile vehicle is available with a full range of implements and should prove a useful vehicle on both larger and smaller acreages, either as the sole vehicle or as a supplementary 'work horse'. (*By kind permission of Duffield ATV, Salisbury.*)

interchangeable simply by undoing four bolts. These sprayers and broadcasters have added versatility in that they can also be fitted to a standard tractor three-point linkage.

The standard sprayer is of a 300 litre (66 gallon) capacity with an 8 m boom and a 400 litre (88 gallon) tank and 6, 10 or 12 m booms are also available. Controlled droplet application (CDA) which reduces the amount of water used, is also available.

The fertilizer spreader is a custom-designed 'Vicon' wagtail broadcaster, fitted with a useful sight-glass to show the amount of fertilizer remaining in the hopper (which is of a 250 kg – 5 cwt or 5 sack – capacity) and an extended on/off handle for operation from the driver's riding position. An alternative 400 kg (8 cwt) hopper version is also available, and a hopper cover which is useful in wet conditions.

Honda ATVs can also supply a more traditional spinning-disc broadcaster either with standard (harder) wheels or a specially extended axle and LGP wheels.

Something which might be of interest on larger farms is the onboard electro-broadcaster, which can be used (with a special mounting bracket for easy height adjustment) for spreading seed or fertilizers. Duffields of Salisbury (the specialist agricultural ATV suppliers) suggest that this, with its 50 kg (1 cwt) capacity hopper, is especially useful for applying fertilizer on hills, or for broadcasting grass seed or slug pellets when the ground is too wet to carry standard machinery.

The onboard agrochemical sprayer mentioned above (see plate 27) is available in several sizes and can be used either with a 6 m spray boom, or a hand lance which would be very useful for spot treatment of weeds in paddocks and around buildings or on drive ways.

A variety of trailers are available, along with a rotary grass cutter designed to cope with heavy growth, with an adjustable cutting height from 1–15 cm ($\frac{1}{2}$ to 6 in) and even gang mowers for use on lawns. Three different types of chain harrows and a half tonne flat roller are also available, and new accessories are constantly being developed. The larger Honda ATVs can be registered as agricultural vehicles for road travel. It is advisable to wear a motorbike helmet, especially on the roads or difficult terrain.

These machines are an exciting development which is likely to catch on in a big way for horse enterprises, and the main UK suppliers, Duffield ATV of Salisbury, Wiltshire, who also sell Suzuki ATVs, are experienced and well able to advise on the most appropriate and economical combination of vehicle and tackle for a given horse enterprise.

Second-hand vehicles are available although, when new, they are economically priced and should be well within the reach of the professional horse breeder/keeper, and worthy of consideration for the amateur horse owner with two hectares or more. The three-wheelers are light enough to lift into a van or Land Rover for transportation and all will fit into a horse box for use as a run-around at competitions. If you do this, do be responsible and considerate of other competitors and spectators, and the organisers and landowners. However, they do far less damage than trials bikes which some competitors use, but could easily be banned if used irresponsibly.

Using your horse or pony for field work

Though complicated jobs such as hay making (see plate 28) may be beyond the scope of most horse-owners, it is worth considering that simple jobs like harrowing can easily be performed, with a little training of horse and handler (fig. 21). Not only may this be beneficial for the paddock, but also for the horse. Some light field work can give a new interest in life to a retired horse or pony, provided things are taken steadily and regularly so the animal does not become sore from the tack

Plate 28 There is no reason why riding horses should not be used for light field work, in this case pulling a hay rake. This one belongs to trainer Paul Weier, at his Equestrian Centre at Elgg, near Zurich in Switzerland. (*Photograph taken by Hans Zuppinger.*)

or tired from using little-worked muscles. It can be a useful fittening exercise for young horses, and even for brood mares in the earlier stages of pregnancy. Obviously the horse or pony used should be of a reasonably sensible demeanour, and should be accustomed to any new activity carefully and gradually.

The act of dragging or pulling something helps, in moderation, to strengthen the *longissimus dorsi* muscles in the back. This is beneficial to both riding and driving horses, and also to broodmares which often develop a sagging or dipped back, often unnecessarily, as they become older due to the strain of carrying the foal effectively 'suspended' from the spine.

A suitable collar or breast collar will be required, and of course a set of harrows. It is best to train the horse initially to pull an old tyre, and then a tree branch or log, but be prepared to release it quickly with a knife or quick release knot/mechanism in case of trouble. The horse will have to be led, certainly at first.

Fig. 21.

It may be as well if you seriously want to use 'horse power' in your fields to attend an Agricultural Training Board course in the subject. They will also be able to advise about suitable tack and implements.

Using Shanks's pony (!)

There is much that can be done to improve pasture productivity by hand, and a number of simple implements are available for this purpose, which may have a place on even the largest units. They are all adaptable and useful for managing awkward corners and odd-shaped paddocks. The use of a fork or shovel to dig up poisonous weeds such as ragwort and for collecting droppings is obvious, and a wheelbarrow is always useful.

Two items which I consider indispensable, especially for small acreages, but always useful for tidying odd corners on a large establishment, are a sickle and a Turk scythe (fig. 22). Neither are suitable for use by children. The Turk scythe is used with the operator more-or-less upright, and once you get 'into the swing of it', it is simple to use. It is light in weight, weighing around 1.5 kg complete with its long 1.9 m (5 ft 6 in) handle. It can be used for 'topping' rough areas in small paddocks, and especially cutting down tall weeds and brambles (the latter with the special bramble blade).

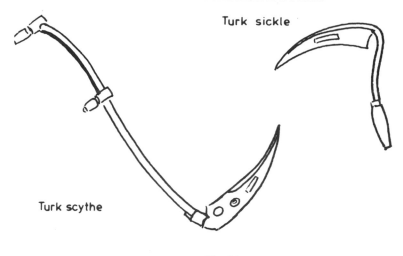

Fig. 22.

You cannot eradicate tough weeds like docks and nettles by cutting, but can stimulate regrowth rendering the plant more susceptible to herbicides if you are spraying late, and of course if you cut them before seeding you can reduce their spread. The scythe is useful around verges and buildings for 'tidying up' purposes, and cannot be recommended too highly.

The same manufacturers also produce a useful hand sickle for odd jobs, but the scythe eliminates the 'back-breaking' aspect of such work where suitable.

Another useful means of cutting down weeds and long grass in odd corners and around fences and buildings is to use a heavy-duty petrol-driven strimmer, a relatively recent innovation which cuts by virtue of a rapidly rotating nylon cord. The strimmer may be of particular interest to horse owners reluctant to use herbicides.

There are a number of ways to eradicate weeds chemically 'by hand', and the hand-held version of the 'weed wiper' is worthy of consideration. The basic principle of both the hand-held and tractor-mounted machines is that a contact herbicide such as glyphosate (e.g. 'Round-up') is placed in a reservoir and filters on to rollers or rope wicks. These are used to 'wipe' the weeds which are growing above the crop or pasture, which gradually die (plate 29).

Weed wipers are best used in warm conditions when they will have a rapid effect on the plants treated and the hand-held version, available from agricultural merchants and garden centres, is particularly versatile. An advantage of this system is that the herbicide is not applied to the grass itself, which is not then 'checked' in its growth by such treatment, which is particularly useful in fields shut up for conservation. Animals should not, however, be allowed to graze until the plants have completely died back, from seven to fourteen days depending on the weather and maturity

Plate 29 The weed wiper – a useful means of controlling weeds on small areas. (*Photograph supplied by Hectaspan Ltd.*)

of the weeds, and the treatment should not be used for ragwort, which becomes palatable when wilted and must be dug up and removed altogether.

Environmentally, wick and roller herbicide applicators reduce the dangers of herbicides being carried on to crops and plants other than those intended (e.g. wind-blown sprays) and of effluent and unused spray getting into water courses as far less water is used. This will also be an advantage where water supply to fields is a problem. If sprays are used in fields with water troughs, these must be covered during spraying and emptied and cleaned before further use, and ponds or slow-moving streams fenced off, at a safe distance.

The other means of applying herbicides to small areas is the use of a knapsack sprayer, which may be purchased or hired from an agricultural merchant or garden centre. These may be fitted with a single spray nozzle, or even a small, hand-held boom or lance with several spray nozzles.

The usual safety precautions for using and mixing pesticides of course apply, and I personally prefer to wear gloves, an overall and a face mask when spraying *any* pesticide, whether the package instructions say this is necessary or not. Spraying should not be attempted on a windy day, and most herbicides work best in sunny weather, after rainfall, when the weeds are growing vigorously. Incidentally, some people find pesticide terminology confusing. Basically, herbicides kill plants, insecticides kill insect pests and fungicides kill mould and fungal infections. All three are classified as pesticides as they all kill pests!

Examples of the knapsack sprayer are the Soho Jet Pack and the Cooper Pegler range. The Jetpack 425 has a tank capacity of 15 litres (3.3 gallons) and weighs (empty) some 4.3 kg ($9\frac{1}{2}$ lb). The tank is moulded on to a strong tubular steel frame and the contoured back and padded carrying straps make it reasonably comfortable to use. The pump design allows applications at pressures of 0–6 atm (0–90 psi). The 425 is operated by a simple hand pump, although the Jetpack 410 and 423 are motorised knapsack sprayers and may be more suitable for larger areas.

The Cooper Pegler CP3 is a 20 litre (4.4 gallon) capacity hand-pump sprayer, and comes with a range of useful accessories. Cooper Pegler also make trailer units for pulling by hand (not on rough going!) and mini-tractors, available in petrol or electrically operated versions.

Knapsack sprayers can also be usefully employed for pest control in stables, feed stores and barns.

Another useful 'hand-held' piece of equipment for field maintenance is the Cyclone hand seeder, available from Hunters' of Chester, the grass seed house. It is a portable seed (and fertilizer) broadcaster, based on the concept of the old fiddle seeder used in days gone by. The seed fiddle has been described as 'something like a cross between bagpipes and violin', a bow was used to spin a disc on to which the seed fell. Vanes on the disc threw the seed out in an arc to broadcast the seed evenly.

The Cyclone hand seeder, as a modern development of the fiddle seeder, has the bow and leather thong that operated the rotary spinner replaced by a handle with a gear which drives a saucer-like wheel with vanes in it. The seed falls from the hopper onto this polythene saucer, and as the handle is turned the two gears cause the saucer to whizz round and throw the seed out. The hopper is designed to hold 4.5 kg (10 lb) of seed and may also be used to spread pelleted or prilled fertilizer.

At a cost of £48 + VAT (1986) this could be an extremely useful tool for patching up with herbage seed on bare spots, and over small areas for both seed and fertilizer. Apparently it is possible to treat approximately 0.4 ha (1 acre) overall in an hour.

There are so many types and permutations of machinery on the market, it would be impossible to describe them all here or offer more specific recommendations. It is to be hoped this section has helped to familiarise horse owners without a knowledge of agricultural machinery with some of the broad principles of operation and types of machine which are available, and give them confidence to consider using these machines for field maintenance and pasture management.

For the larger units who wish to employ primarily horse-trained staff, it may be a good idea to select those with a National Certificate in Management of Horses (NCMH), who apart from studying horse and stud management are required to pass examinations and practical tests in agricultural subjects including tractor driving and basic machinery maintenance, as well as husbandry of beef and sheep which may prove useful on mixed enterprises. At least people qualified in such a way have a useful grounding in such subjects.

For unqualified staff on the stud or horse farm, the use of day-release courses at local agricultural colleges, manufacturers' training sessions and Agricultural Training Board courses (there's even one on using horses on the farm) should be used to the full to give staff adequate training in machinery handling and maintenance, to enable them to do a reasonable job of field operations.

Without trained staff, specialist operations such as baling, sowing, ploughing, etc., should where necessary be left to a suitable contractor. It is to be hoped this book will enable you to make the contractor aware of your specific requirements with regard to horse pastures.

A final word about safety. Apart from removing horses from paddocks treated with sprays or fertilizers for the required period, there is the question of leaving horses in paddocks whilst topping operations, hedging, spreading calcified seaweed, harrowing, etc., are performed. In this context, it could be said that it is a good education for them to become accustomed to such machines, which is true, but *not*, I would suggest, in the same field, unless there are trained personnel there to supervise. Do not leave a non-horsey contractor merrily topping a field containing horses and expect him to cope if they become upset.

A recent example is of a field being topped, containing two horses who were ignoring the tractor. Unfortunately, a sudden rain squall set the horses galloping, they were spooked by the tractor in their path and blinded by the rain, and one promising youngster staked herself badly on a metal fence post. The contractor, who sensibly phoned the owner, was mortified, but I believe should never have been put in that frightening situation in the first place.

Familiarity can breed contempt and I have also seen a young horse, quite at home with a tractor in the field, who, in ignoring it, gaily galloped over the spiked chain harrows it was pulling, with horrific results.

Yes, leave horses in adjacent fields under supervision to accustom them to sights and sounds, but I would say *never* leave even the ploddiest, most 'bomb-proof' old pony in the same field as machinery being operated. It simply isn't worth the risk.

Forage conservation and handling machinery

The pros and cons of different types of forage cutting, conditioning, collecting/baling, stacking/storing and drying machinery are really beyond the scope of this book. Suffice to say that whatever the system you choose, you are, to a greater or lesser degree, at the mercy of the weather, and the object must be to get the crop in as quickly as possible to obtain maximum nutritional and physical quality.

If possible, silage should be cut and left to wilt in the swath before collection with the forage harvester (fig. 23), to reduce moisture content and effluent (in clamp silage), and especially if it is to be baled and bagged. In the case of hay, a modern mower with built-in conditioner may be used, or the crop may be mown and then conditioned, e.g. with a crimper which squashes the stems so that they dry faster. However, overconditioning can cause the leaves to dry out too much and they may then shatter and be dropped or crumble to dust, giving dusty hay. As most of the digestible nutrients are in the leaves this is obviously undesirable.

Once the crop is cut, and if desired conditioned, the swaths should be turned frequently to aid drying on all sides, and rowed up at night (two

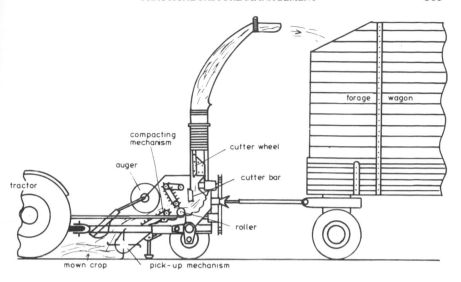

Fig. 23 Precision chop forage harvester

swaths rolled into one) to reduce the surface area exposed to dew or rain, or if rain is expected. Rain soaked or sun-bleached ('mow-burnt') hay will lose much of its soluble nutrient and vitamin content and may be of little feed value. Plates 30 and 31 illustrate machinery to aid drying of the hay crop.

Once the crop is ready it is baled as rapidly as possible. If baled wet it may be inclined to heat in the stack and may even spontaneously ignite, and at the very least is likely to go mouldy and become unfit to feed, especially if a preservative is not applied.

Plate 30 A 'hay bob' used to aid rapid drying of the hay crop.

Plate 31 A spider-wheel tedder, used to turn the hay crop to aid drying, and row up two swaths into one to reduce rain or dew damage and for baling.

I personally feel that in ninety-five cases out of hundred in the UK, the best hay will be made by barn drying it, thus reducing dependence on the weather. This can be done even on quite a small scale in a field stack using a portable fan run off a tractor PTO or batch drying methods. (See *Horse Feeding and Nutrition* by this author, published by David and Charles).

If you are going to use a preservative it may be applied as you bale the crop by an applicator fitted to the baler. Figure 24 shows just how much rough treatment the hay receives as it goes through the baler, which can result in considerable losses through leaf shatter if the leaves are too dry. It is best not to make bales too dense, especially if they are to be handled for feeding by children, although a tighter-packed bale is needed for barn drying, as the wire/twine will slacken off as moisture and therefore volume is lost.

Round bales are literally rolled up in the baler, and this should be set to make smaller bales than for straw as they will otherwise be too heavy. Specialised equipment is needed for handling large bales, but they can be useful for feeding groups of horses in sheds as you just need to start the bale off and unroll it all in one go. Bale collecting and handling equipment is really a matter of scale and convenience and beyond the scope of this book.

A rough estimate of the amount of hay in a stack is given by:

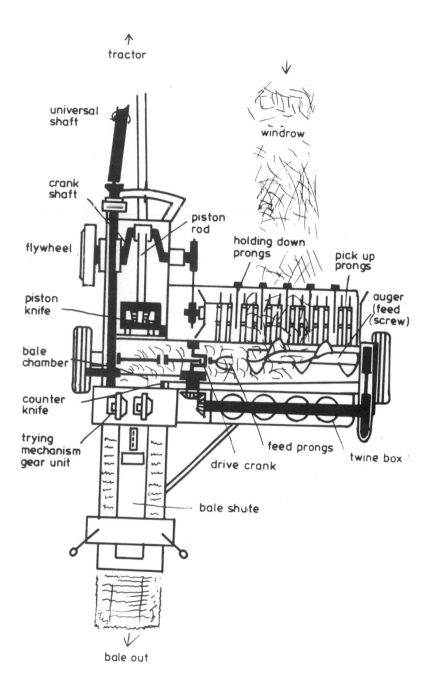

Fig. 24 A traditional pick-up baler.

$$\text{Approximate weight of hay in stack (tonnes)} = \frac{\text{height (m)} \times \text{breadth (m)} \times \text{length (m)}}{11}$$

$$\text{For a round stack} = \frac{3\tfrac{1}{7} \times \text{radius}^2 \text{ (m)} \times \text{height (m)}}{11}$$

(For straw, put both over 16.)

Hedges and trees

In general, all horse pastures will benefit in one way or another from the presence of hedges and trees, although it must be said that even the stoutest hedge in the face of the prevailing wind is no substitute for a properly constructed field shelter.

To be of any value as a wind-break, hedges require some judicious management to prevent them getting thin and straggly, and the attentions of a professional hedge layer will pay dividends. Once the hedge is properly laid, it can be machine trimmed on a regular basis.

However stout the hedge, it is always difficult to guarantee it stock-proof, and it is always a good idea to line hedges with a good fence, especially for field boundaries which give access to roads or potentially poisonous crops – i.e. most hedges! Ditches should also be fenced off, and it is important to ensure that foals or small ponies cannot roll under the fence, or break through a hedge, into the ditch.

Obviously newly planted hedges will require good protection from stock, and it is a good idea to take expert advice on varieties of hedge plant to choose for your soil type and location. Obviously poisonous plants (eg. box, yew, privet, rhododendron, laurel) should be avoided. I have recently visited a thoroughbred stud where the stallion paddocks have been hedged with laurel. This is asking for trouble. Blackthorn, hawthorn, may and beech are all suitable hedgerow plants. They say with an old established hedge that you can tell the age of that hedge by a hundred years for every species of hedge plant that is established in it.

For those horse keepers who are interested in conservation, the following notes, adapted from an article which appeared in *Farmer's Weekly*, 27 September 1985, about management of hedges for game may be of interest:

'More than 90 per cent of the country's partridges and up to 50 per cent of the pheasant population are thought to nest in hedges. But research by the Game Conservancy has shown that almost half the hedges in Britain are no longer suitable as nesting cover – because they are managed incorrectly. A simple change in the frequency and timing of hedge trimming is often all that is needed to greatly improve nesting cover quality.

The best type of nesting hedge will lie within a field boundary 3 m to 4 m (10 ft to 13 ft) wide, although the hedge itself should be no more than 2 m (6 ft 6 in) wide. On each side of the hedge a raised earth bank should be created (where these do not already exist) and maintained to provide a grassy verge to the field boundary. Such banks are favoured nesting sites for partridges and help reduce the likelihood of nest predation.

Grassy banks are best for game when they are covered with suitable vegetation at the correct time of year. Birds begin looking for nest sites late in winter – in some years as early as January – when hedge bottoms are often laid open and bare. But if 25 to 40 per cent of the base area of the field boundary has a covering of residual dead grass, the hedgerow will be close to ideal for nesting.

Such vegetation is particularly important for the grey partridge which covers its eggs with dead grass to protect them from predators. Because of this, nest losses during egg laying are much lower for our native partridge than for the red-legged partridge and the pheasant, both of which were introduced to Britain.

To maintain grassy banks and good nesting hedges, trimming should be done only once every other year. This not only creates good nesting cover for gamebirds, but provides plenty of berries for songbirds in winter. Mechanical hedge cutting is not in itself a problem, provided neither the hedge nor the verge is cut back too hard and trimming is carried out in September so some late season growth occurs, giving vital protection the following January.

The role of the hedge is to provide protection from both the weather and predators. The tall, neglected hedgerows that are common around farm and parish boundaries are invariably too empty at the bottom to offer good cover for nesting game. At the same time they often support an abundance of predatory birds and mammals. A row of narrow stumps is no better.' Such hedgerows are also likely to encourage rabbits whose warrens may be dangerous to horses and are to be discouraged on horse pastures. 'The best hedges are of good solid thorn, between 1 m and 2.5 m (3 ft to 8 ft) tall and 2 m (6 ft 6 in) wide. When combined with grassy banks on either side, this type of field boundary is as near ideal a nesting habitat as British game birds are ever likely to find ...'

I am very much in favour of conserving wild plants and flowers, both for themselves and as a habitat for butterflies and other insects, and small birds and mammals. However, be aware that shady hedge bottoms are often a habitat for poisonous plants such as wild arum, mare's tail, foxglove and so on, and that deadly nightshade and bryony are often found in the hedge itself. These must be eradicated from horse paddocks. If you want to preserve them, earmark a corner somewhere to 'go wild' and make it *absolutely* stock-proof. This may cause a howl of anguish from some conservationists – rabbit runs and so on are all very well, but a horse which might break its leg galloping over a rabbit warren also needs 'conserving' and a suitable balance must be struck.

Do not be tempted to use proprietary 'conservation mixtures' of seed, as they often contain poisonous plants. If you want to add variety, use edible herb mixtures. Again consult an expert who understands conservation as a *balanced* aspect of farm crop and animal management (see below).

With regard to trees, a managed copse can provide a useful source of income as well as a habitat for wildlife and a wind break for your pastures. Single or rows or groups of trees can provide welcome shade from the hot summer sun, both in field boundaries and grouped or singly in the pasture itself. However a number of points should be taken into consideration:

(a) No poisonous varieties, either of the tree bark/leaves or seeds it drops (eg. horse chestnuts (conkers) and acorns can be poisonous). Even if a large tree is out of reach of horses, a bough may be blown down in high winds, into the paddock, and the wilted leaves may be both tasty and toxic. By all means preserve trees where you can, but lop off over-hanging boughs, or fence well round the whole spread of the tree (with foal-proof fences) if necessary. Make such fences in field boundaries graduated, with no sudden projections into the field which galloping horses may injure themselves on or bullied ones be cornered in.

(b) Small fields surrounded by trees may be very shady and stay wet for a long period and some 'thinning out' may be desirable.

(c) Horses who stand for long periods under dripping trees can be especially subject to 'rain scald', as the water drips on to their backs and a small microbe related to the one implicated in mud fever on the legs causes the skin on the loins and croup to become sore and cracked and suppurate. This condition *can* be fatal, like mud fever, if septicaemia sets in.

(d) Water troughs should not be sited under deciduous trees as dead leaves will accumulate and 'sour' the water.

(e) Chewing tree bark is probably not good for the horse, and certainly not good for the tree. It may be indicative of boredom (a crib-biter), lack of fibre in the diet, upset digestive micro-organisms, or lack of minerals, and whilst protecting the trees, efforts should be made to find the reason for this behaviour and eliminate it (e.g. feed straw for fibre, probiotics to restabilise the gut environment, or a mineral supplement if required). For crib-biters, muzzles and cribbing straps only mask the symptoms, they do not cure the problem by removing the cause.

(f) Obviously young trees must be protected from horses.

Again, there are experts who can advise you on conservation and management of existing trees/copses/woods, and on selection, planting and management of new trees.

The British Institute of Agricultural Consultants has a number of members around the country who can advise on hedges, trees and woodland and general conservation matters on farms and estates, and their members may be contacted through the secretary whose address appears at the end of this book.

Fencing, man-made wind breaks and gates

The availability of materials for fencing and so forth will, to a certain extent, dictate the choice of a particular system. Obviously economy is another major consideration, but suffice to say that it is always preferable to obtain the best you can afford, in terms of safety, longevity and ease of maintenance.

Fencing

Apart from considerations of cost and appearance, the *major* considerations when selecting fencing must be safety and suitability for the stock in question. The fencing selected must be of a type which is unlikely itself to inflict injury, and which is proof against the stock which will be contained within it. It should be high enough at the top to deter stock from jumping out, or fighting over it, and low enough at the bottom to prevent foals and small ponies from rolling or crawling underneath and getting out, or in the former case, becoming distressed and injuring themselves when they find they cannot get back to their mothers, or drowning in a ditch the other side of the fence. The fence should also discourage animals from leaning through to graze and/or pawing and getting a foot caught.

When laying out fields, the siting of gates and water troughs deserves some consideration, and on studs it is almost always desirable to use double fences between paddocks to discourage fighting over fences, and it may be convenient to have a stout hedge between these for use as a wind break, or even to grow comfrey in the gaps for convenient cutting and dropping over the fences on each side for the stock. Projections such as trees and tree roots should be fenced gradually and sudden projections in the fence avoided.

Where foals or youngstock are kept, it is also a good idea to round the corners of the field, preferably by actually rounding the corners, or if not by fencing diagonally across square corners, leaving a triangle which can be a useful place to site a tree or comfrey patch. The object of this is two-fold. Firstly, the curve will gently guide galloping youngstock, and brood mares whose foals have a tendency to gallop 'blind' at their side and can collide with the fence or the mare if the mare suddenly turns around the corner (plate 32); and secondly to prevent animals from being bullied and cornered in kicking matches.

Plate 32 Curving the corners of stud paddocks can prevent accidents by gently guiding galloping youngstock round the corner. (*By kind permission of Mrs Dare Wigan.*)

If foals are to be creep fed, a safe area which is inaccessible to greedy mares should be provided (fig. 25). There are many ways of doing this, but safety and prevention of cornering or bullying should always be considered (plate 33). If there are only one or two foals, the triangle at the corner of a squared-off field may be used, by not quite taking the diagonal

Fig. 25.

to the boundary *at both ends*, so that there is a separate entrance *and* exit. For larger numbers of foals this may not be suitable as there are too many corners for bullying so a larger area adjacent to a fence, or even better still a circular creep area out in the field with several exits and entrances should be constructed. The disadvantage of this latter is that it is necessary to enter the field to put in the creep feed. Water should also be provided in a creep area, in a container low enough for the foals to see and reach into, and stable enough not to be tipped over.

Mares may be prevented from entering the area by making gaps for foals too narrow or too low for them. This can be a problem if pony mares and their foals are pastured with larger ones. When placing feed in the creep area, site it well out of reach of the mares to discourage them from leaning on the fences to try and reach it, and provide several containers if there are a number of foals, some distance apart. Keep all feed and water containers scrupulously clean. Mares should of course receive adequate

Plate 33 The foal may be encouraged to eat solid food by its mother's example, but should be provided with a creep manger or area so greedy mum doesn't eat the lot. Alternatively a larger trough which they can both eat from would be preferable to this little bucket. (*Photograph taken by Terry F. McCarthy.*)

hard feed whilst lactating, and it is cruel to creep feed foals if mares are hungry and subsequent jealousy may lead to accidents.

Stallion paddocks will require higher fencing, double, and preferably sited a safe distance from other stallions and in-season mares, unless of course they are for use as teasing paddocks. Fences for thoroughbred stallions should be 1.6–2 m high (5–6 feet), and an alleyway at least 3.2 m (10 feet) wide should separate them from other paddocks. Some breeders also run electrically-charged wires along the top of a stallion's fence to reduce aggressive behaviour over the fence.

Paddocks for thoroughbred mares and youngstock will require fences of no more than 1.5 m (4 feet 6 in) height. If paddocks are side-by-side, and especially when not separated (visibly) by a hedge, or with square corners, it is desirable that they be of similar length, so that youngstock concentrating on playing and galloping together along a fence on both sides do not have accidents as the ones in the longer paddocks gallop on, while the others crash into the fence at the end of a shorter paddock.

Post and rail fencing

The most desirable fencing must still be of the post and rail type, which may be made of a number of materials. The obvious one is *wood*, and in order of preference I would take (1) all hardwood, (2) hardwood uprights with softwood rails, and (3) all softwood. All wood should be treated with a non-toxic preservative to enhance its life, especially the uprights which go into the ground. These should be well-tamped down to reduce access of air and water which encourage rotting.

Wooden fencing is attractive, highly visible and durable, but it can be susceptible to damage by crib-biters, especially if they are on an unbalanced diet. It has also been suggested that wood which has been preserved by tanalising or tanerising may encourage crib-biting because tanalith contains salt.

Fencing may be painted or creosoted, but both are labour intensive, and it has been suggested that creosote may be carcinogenic (ie. predisposing to or causing cancer). Certainly animals should not have access to fences treated in either way until the paint or creosote is dry, and if paint is applied by spray gun and has coated excessively grass at the base of the fence, it is probably best to mow and remove the coated grass.

Personally, except perhaps adjacent to buildings, I think miles and miles of white-painted fencing looks frightful, but then I'm rather fond of nature and like my wood to be wood-coloured! However painting does enhance visibility and it may be sufficient just to paint the top rail. Do not paint the underside of the rails as this prevents any moisture from escaping and can accelerate rotting.

Recent studies indicate that unpainted oak fencing may last up to twice as long as painted oak, and also that the heartwood cut from the centre of a tree is more durable than the slabwood or sapwood from the outer layers.

Other preservation alternatives are to apply asphalt-based paint to the in-ground sections of fence posts, or buy posts treated with a preservative solution of copper, chromium and arsenic (CCA). This preservative is useful around buildings as it is also flame-retardant. Pine and other softwoods treated in this way are easily identifiable by their greenish appearance. They will last up to twice as long as the untreated wood but cost around 33% more, so, in the long term because you only have to buy one set of more expensive fence instead of an initial and replacement set of the cheaper fence, the former is more economical.

Other materials used as wood preservatives include tar and used engine oil (which may contain lead) and neither is suitable for wood in horse paddocks or buildings. Fence off telegraph poles which have been dipped in tar. Wood preserved in arsenic is deemed safe as the chemical is inert when the wood is intact, but be aware that it releases the poison when burned rendering smoke and ashes toxic.

Post and rail fence may be of two types. The rails may either be nailed onto the uprights, in which case they should be nailed on the inside of the field, so that animals leaning on them do not dislodge them, or they may be morticed, in which case the rails are inserted into holes in the posts. Sometimes a staggered combination is used, whereby the rails are nailed to posts at each end and pass through a central one.

Posts should be placed with the wider face in line with the length of the fence, giving a larger surface for nailing and providing a greater resistance to overturning. Where the rails are morticed it is necessary to provide a sufficiently wide face on the posts for the mortices, and it is usual to place the posts with the wider face at right-angles to the fence.

Posts 2 m in length are suitable for a three-rail fence 1.2 m high. If a higher fence is required, then longer posts must be used, rather than gaining height at the expense of depth below ground level. The minimum dimension suitable for rails for either kind of fencing is 75 × 38 mm sawn timber. Half-round rails may be used for a nailed fence, two rails being cut from a barked and peeled log of not less than 100 mm diameter. For nailed fences the ends should be butted and not spliced. Galvanised nails should always be used. Rails should span three posts and be staggered in such a way that all the joints in a run of fencing do not fall on any one post. Posts should therefore be more than 1.8 m apart.

A variation is the Sussex fence, in which roughly hewn posts and rails are used. The ends of the rails are tapered to fit into mortices cut side by side in the posts giving a slight zig-zag pattern to the fence.

Ideally for horses, the top rail, of whatever material, of a fence should be at the top of the post. If the post protrudes above the top rail, injuries can occur if horses bang their chins down onto the fence. It is often advisable to slant the top of the support posts slightly to encourage rain water to run off.

It may be worth considering that in the event of accidents, oak tends to form long and dangerous splinters whereas pinewood breaks more cleanly, which may be an argument for using hardwood uprights but softwood rails. The former are especially desirable in wet or marshy land from a durability point of view.

Pipe fencing on a post and rail system is also favoured by some studs. It is relatively safe, very durable, but also expensive and requires regular painting.

A promising development in the USA is solid plastic post and rail fencing such as 'Saratoga fence' which is described as being maintenance-free and does not chip, crack, rot, rust or peel, and never needs painting. It is white, so highly visible. This brand carries a twenty year warranty so presumably lives up to its reputation, though I have no experience of its use. It apparently contains additives to enhance impact resistance and ultraviolet inhibitors, presumably to prevent cracking, and stands up well to winter weather – it sounds excellent, so long as you like white fences.

Another new development is to have a post and rail fence with wooden/metal/plastic posts, and flexible PVC or rubber-coated webbing 'rails'. The rubber–nylon variety (e.g. cut from strips of conveyor belt material) has the significant drawback that horses, especially youngsters, tend to chew it and may suffer from colic or impaction from swallowing the rubber or cords it contains. Yards of cord have been found in the intestines of some youngsters at post mortem in the USA. However, it is cheap, sturdy, resilient and impervious to weathering, and reasonably attractive.

'Flexafence' is a British development along these lines. It is made from a specially woven and treated nylon webbing covered with non-toxic PVC and is nailed on to wooden posts.

Another alternative, also developed in the UK is 'Studrail', which combines cores of high tensile wire with a smooth soft PVC covering. The idea is that the inherent elasticity of the 'rail' will absorb impact and gently rebound without splintering or shattering. However, a number of stud users, especially those with cattle as well, have found that it does tend to stretch somewhat over a period.

Studrail has the advantage that it is available in white, black, brown or green, and other colours to order, and it is supplied in 100 m coils of 75 or 100 cm (3 or 4 in) widths. Despite the reported stretching problem, this may be of particular use for private paddocks, rather perhaps than those subjected to heavy and intensive use, and there are good reports of it from private horse owners. The Studrail may also usefully be combined, as a top rail, with wire mesh, or as an eyeline above wire strands.

Apart from modern alternative rails, there are some new ideas around about posts. For example, 'Metpost steel post supports' (UK) comprise a square-section heavy gauge steel socket with a long, tapered metal spike. A report in *Riding Magazine* (July 1986) by Jay Swallow describes how the spike is placed where the centre of the post is required, and hammered home until it sits flush with the ground. The fence posts are then tapped into the socket and held firm by wedge grips. The system prevents rotting of the posts and saves on post length. Rails or wire can be fixed to the posts. It strikes me that anything which can be as easy to assemble may also be easy for a horse to demolish, so check with existing users before installing this system.

In the USA Fox Valley Draft Horse Farms have developed the 'Shire Horse Fence Clip' which is an 18 gauge galvanised steel connector designed to slide over steel fence posts and hold two 25 mm × 150 mm (1 in × 6 in) finished boards overlapped. They point out that they are strong enough to hold their shire horses, but I feel that metal fence posts and younger boisterous horses do not in general mix very well.

Wire mesh fencing

Woven V-mesh wire fences, properly erected and maintained, can be useful for horse paddocks, especially where foals or sheep have to be contained. The small V-shaped openings prevent a horse from getting its foot caught, and the V-mesh configuration is extremely strong and durable. Unfortunately, mesh wire can become bowed out by horses rubbing or pushing on it and consequently it is desirable to place a pipe or rail across the top and at mid-height to increase visibility and add support.

Chain-link fencing is not really ideal for horse paddocks as the twisted ends are very sharp, so it is necessary to bury the bottom of the fence in concrete, and again there should be a pipe or board at the top to protect horses against the sharp points, and increase visibility.

Stud Fence by the makers of Studrail mentioned above (Animart, in the UK) is a meshed high tensile wire on eight foot strainers, eight foot struts and seven foot posts all in soft celurised wood. It stands 1.2 m high and is best used with a run of Studrail above, for enhanced visibility. The mesh is designed so that the gaps at the bottom are only 7.5 cm high, graduating to 20 cm high at the top. This means that if a horse paws the fence there is no chance of getting his foot caught. The wider spacings higher up mean you are spared the expense of a close mesh throughout.

This standard Stud Fence is intended mainly for use as perimeter fencing, and Animart recommend that their special close-mesh Stud Fence be used to divide fields where horses may be grazing on either side. In this case the smaller 7.5 cm mesh continues to the top to safeguard against legs being trapped if a horse should kick out whilst squabbling with its neighbours. All Animart fencing carries a twelve year guarantee.

It is vital to get the correct tension and straining arrangements when erecting any form of wire fencing. Plate 34 illustrates a particularly poor form of fencing.

Plain-wire fencing

I do not like using wire in horse paddocks but there are always going to be occasions when this is deemed unavoidable. Ideally, it should *only* be used with a top rail or pipe to enhance visibility, and tension *must* be maintained at all times. The only sort I would feel reasonably happy about is the highly tensioned post and dropper type using high tensile

Plate 34 A vivid illustration of the pitfalls of sheep netting, combined with barbed wire and a bare field. If the horse doesn't get its foot caught it has a good chance of tearing its chest on the barbed wire. The wire is very loose, the fence posts look flimsy, there is no shade from glaring sun or shelter from driving rain (the bushes are some way from the fence), and barely any grass at all. Sadly this is a typical pony paddock.

wire. Metal posts are generally used to aid tensioning, and straining posts may be alternated with plain posts. The function of the droppers is to keep the wires correctly spaced.

The method of tensioning or straining is very important. Eye bolts, hook bolts or winding brackets can be used on wooden, concrete or steel posts, although the raidisseur type is generally used on steel posts where there is little bearing for eye or hook bolts.

If the flames from heather or straw fires are allowed to reach a high tensile fence, the wires may become brittle and snap.

Barbed wire fencing

Horses and barbed wire quite simply *do not* mix, no matter how well-constructed a barbed-wire fence may be in agricultural terms (see plate 34). I have heard it said that barbed wire is better for horses than plain wire because they know when they've hit it! That smacks of 'shutting the stable door when the horse has bolted', as you and your vet will also probably be well aware that the horse has hit it.

If you *must* use it keep it for the top row of a plain wire fence, and only for field boundaries with a hedge behind them to discourage the horse from pushing through (although it may then lean over to eat the hedge – you can't win with barbed wire!). *Never* use it to *divide* fields containing horses as fights over the fence and jumping out can result in vicious injuries.

Festoons of barbed wire, metal posts leaning at drunken angles, bits of string and a few broken pallets are asking for injuries. Injuries in that situation are *inevitable, not* accidental.

The worst barbed-wire injuries I have seen involved two ponies in a small triangular field with straw being burned on two sides and a dual carriageway on the other. The terrified animals were trying to escape the flames and became thoroughly tangled in the fence. *Always* ask neighbouring farmers to warn you if they are burning straw or stubble so you can move livestock.

Electric fencing

Like barbed wire, single-strand electric fencing and horses do not really mix, mainly because galloping horses simply don't see it. However, there are alternatives which may be used, always within a more sturdily fenced field boundary to be sure of preventing access to roads and other dangers should horses get out, as a means of temporarily dividing up fields, strip grazing, preventing access to hedges and so forth.

One is the new *reflective* electric fencing which consists of a single flat metallic strip, which is twisted as it is put up and glints in the sun making it highly visible. I have no experience of the use of this but it certainly sounds like a useful idea.

The other is an adaptation of *electric sheep netting*, which is now available in the UK in a taller version specifically for horses (112 cm – 44 in). In the UK, this is manufactured by J. & M. Gilbert. The fence comes in kit form, complete with stakes attached (although they can be easily removed if desired), and all you need to supply is a standard electric charger unit. The fence itself consists of a series of electroplastic twines interwoven to make a grid. The posts are white and the horizontal twines black and yellow, and the vertical wires bright yellow, making a solid-looking barrier which should be easily visible to a horse.

Meshed electric fencing is already used extensively in the USA and is catching on in the UK.

Electric fencing may be particularly useful for horse owners who rent fields with sound boundary fences, but wish to divide them up into smaller units without spending a lot of money on someone else's field.

The safest way to accustom a horse to electric fencing is to introduce him under controlled circumstances. Take the horse on a head collar and lead rope, wearing gloves and a crash hat yourself, and let him have a look. He will instinctively stretch out his nose to sniff it and will, of course, get a shock. You can then reassure him and turn him loose yourself, rather than have him experiment for himself and give himself such a fright he panics and gallops off and goes straight through another stretch of fence. Once having sampled the fence, he is unlikely to test it out again.

Badly sited fences and gateways can be a continual nuisance to the day-to-day running of a horse farm. The visual impact on the countryside should also be considered, taking into account that, where possible, it is desirable to avoid fencing along a sky-line and better to follow the contour just below it. Fencing should also, where possible, follow the contour of the land rather than cutting across it.

Where public or other rights-of-way cross the land, stiles, kissing gates, or in the case of bridleways, self-closing gates must be provided for walkers and riders.

Man-made wind breaks

A fairly recent development which may be useful in open country where there are few hedges or dips in the ground to provide shelter against the prevailing wind are wind breaks, which may be sited in the fence line, consisting of a wooden or metal frame up to 4 m (12 feet) high with specially designed plastic strips or mesh stretched across it which 'breaks up' the wind and reduces its force. These can already be seen around farm buildings, protecting glass houses and even the customers of petrol (gas) stations in open countryside!

A 5–10 m stretch may provide welcome relief in areas with strong prevailing winds, and should be sited adjacent to a well-drained spot where poaching will not prove to be much of a problem. A wind break for mid-field erection which caters for changes in wind direction is also available. This would certainly be preferable to a dark little field shelter to many horses and ponies in all but the worst weather conditions, and is well worthy of consideration.

Gates

How well a gate is hung is rather more crucial than the exact design chosen, apart from the fact that gates should be solid, high enough, horse-proof and safe (see plate 35).

Some people prefer tubular metal gates with thick steel mesh with small holes on the lower half to prevent pawing horses getting feet caught; others favour a traditional wooden five-bar gate. For studs I favour tubular metal gates with a frame which is low to the ground and uprights only, so it is difficult for animals to get a leg caught.

Due to their weight, timber gates are rarely more than 3.6 m in width, and require substantial hanging posts set in concrete. Unless they are made of heart-oak or larch, timber gates will need to be preserved against decay, preferably by the use of pressure impregnated timber.

Under no circumstances should gateposts be used to attach strained wire fencing, as this will upset the hanging of the gate. A well-hung gate should be latched easily; open easily with the minimum of effort, to where it will stand on its own accord; and close gently and smoothly, finishing by its own momentum, without hitting the shutting post with a crash.

Security

In this day and age, when horse stealing is on the up and up in view of the high value for meat of even the most ordinary horse it is prudent to:

(a) chain and padlock gates *at both ends* so they cannot be lifted off their hinges,

(b) Freeze-brand horses for ease of identification (a painless procedure), and keep a stock of photographs for immediate circulation to police, horse sales and markets, and

(c) Make friends with people living adjacent to your paddocks and ask them to 'keep an eye open'. A bottle of wine or a box of chocolates each Christmas is a small price to pay for some measure of security.

Give them your telephone number and ask them to contact you/the police if they see anything suspicious and note licence plate numbers, makes and colours of suspicious vehicles. This applies to vandals as well as thieves of course.

Field shelters

Even the hardiest native ponies require some form of natural or man-made shelter when confined by man to fields and no longer able to roam around and seek their own shelter. Most horses and ponies, however, will stand outside even in the most appalling weather conditions rather than enter a dark, pokey shed, as horses are naturally claustrophobic and suspicious of dark enclosed areas.

So, a field shelter should be light and airy, with its back to the prevailing wind. A wide roof overhang at front and sides will encourage animals that don't go in to at least shelter a little, and provide some welcome shade in the summer. The inside corners should ideally be rounded off to prevent bullying and cornering, and any support posts securely padded. Gates may or may not be added to facilitate shutting an animal in, and if they are, a water source must be provided.

Plate 35 Apart from causing an obstruction in the field, this water trough is badly sited – being next to a gate where poaching will already be a problem. It does not have the advantage of serving two fields at once. The gate is not really suitable for horses as there are too many places for a foot to get caught.

Whether or not to provide bedding depends on individual circumstances, but if horses are to be wintered out, a concrete apron will be desirable to prevent poaching of the surrounding area. I have seen one shed for several larger horses and a Shetland pony with an escape door-flap at the back for the Shetland which was rather susceptible to bullying – a sort of giant 'cat-flap' which the pony used readily whether it needed to 'escape' or not! If you intend to do your feeding in the shelter, it must be especially roomy to avoid bullying.

In the UK, buildings under 30 m² may not require planning permission and buildings under 115 m³ may not require building control approval. (The BHS can advise on this.) VAT may not be chargeable if the buildings are erected in a field with no other buildings.

Field shelters may be open-fronted (safest) or part-enclosed and should be constructed on a concrete or a drilled and rodded sleeper base. The roof and roof overhang should be high enough to prevent a horse tossing its head from banging its head or eyes. Confections cobbled together out of corrugated iron and pallets, etc., are usually dangerous and an eyesore, and may be unbearably hot in summer, so are obviously to be avoided.

Water sources and feeding equipment

Water sources

Water troughs should preferably be of the self-filling variety, and should be easy to empty and clean. They may be cleaned by scrubbing with plain water or suitably diluted dairy hypochlorite, which should be rinsed off before refilling.

They should be sited in a well-drained area of the field, and some provision made to reduce excessive poaching, including prevention of leakage and overflowing. They should not be sited near or under deciduous trees as leaves will drop into the water, or near gateways or other areas subject to poaching.

Ideally they should be sited along a fence or recessed into it, rather than at right-angles to, or in front of it (plate 35). If they cannot be in the fence they should be well in front of it, three or four horses' lengths, to allow free access all round and prevent horses becoming stuck behind them. However this does then mean there is an obstruction out in the field. They should be high enough to prevent horses pawing the water, with the underneath filled in to prevent foals getting stuck. If necessary a separate source for foals should be provided, that they can see into and reach easily (plate 36).

Plate 36 This trough is at a convenient height for both mare and foal to drink, but should be blocked underneath to prevent a rolling foal from getting stuck. The fence of alternating wire and rails is rather flimsy, and would be safer if the top rail was flush with the tops of the posts, and the bottom wire replaced by a 'foal-proof' rail.

Any protuberances or taps should be safely boxed in. An old bath with taps sticking out and sharp edges is asking for trouble and simply not adequate. Water buckets as the sole water source are not adequate either, as they are easily tipped over.

It should be remembered that, on average, horses require 45 litres (10 gallons) of water a day *each*, more in hot weather, if they are in work or lactating, or if the pasture is not lush (e.g. high summer). Many horses colic because they are dehydrated in winter because ice on their troughs has not been broken. This should be done twice a day, every day, without fail.

If you do not have water piped to your fields, it is worth considering finding your own water source and sinking a bore-hole by employing the services of a water diviner. This may be a rather mysterious art, but it certainly isn't 'mumbo jumbo'! A good water diviner can tell you the precise location and depth of a water source through several hundred feet of solid rock, and some even operate from helicopters (e.g. to tell mining companies where to *avoid* digging or drilling because of water) with great success!

Pat Lucas, a well-known British equestrian writer, is also an accomplished water diviner and has recently helped to locate a source of underground water for Chepstow Racecourse, who were delighted to be able to provide ample water for watering the course, etc. Pat is based in the Wye Valley and her address, and that of an association of water diviners and dowsers who can give you details of members in your area, and internationally, is also given at the end of this book.

Stagnant ponds and sand- or clay-based streams should be safely fenced off. Although a natural water source sounds nice, it is only suitable if it has a stony or rocky bottom, to avoid sand ingestion and sand colic, and poaching of the approach, and the water should be running and not still (plate 37). Check for possible sources of pollution upstream (e.g. industrial or nuclear plants, silage or slurry effluent and so forth), and if in doubt have the water analysed by your county analyst (UK) or State Extension Service (USA) on a regular basis.

Plate 37 People have romantic notions about using 'natural' water sources but this mud bath with a shallow, silty bottom is a recipe for sand colic.

Feeding equipment

Hay racks should be designed in such a way that they preclude animals getting feet caught or heads banged, avoid hay seeds dropping into their ears and eyes, have no sharp corners or projections and they should be stable, durable and easy to relocate (see plate 14). They should be placed in a well-drained spot, preferably away from the fence, with good clearance all round.

Round cattle feeders should only be used with curved separators and no top bar, and a high enough solid section to prevent legs getting caught (see fig. 12). Hay nets should be tied to a *solid* fence post or tree, high enough so that the smallest animal in the field cannot get its foot caught when the net hangs down empty. Small mesh nets should be used for hayage or MVP (moist vacuum packed) forages to slow feeding rate.

There are pros and cons for feeding hay on the floor, the main disadvantage being wastage. Always allow enough nets or piles of hay for one more horse than there is in the field, and space them four or more horses' lengths apart to prevent bullying, and ensure there is enough feeding space at hay racks.

A new development is the hay tube, a sort of 'flute' you push hay into at the side and the horses eat out of holes in the top, rainwater draining through smaller holes in the bottom. Whether this will work in practice remains to be seen, but the big advantage over nets is that they are easy to fill and don't droop down dangerously when empty.

For hard feed, a non-spill feeder is described elsewhere (page 123), and specially designed bowls may be dropped into tyres to give them stability. If you have solid post and rail fencing, a useful alternative is the type of feeder designed to hook on the fence or over a stable door (plate 38).

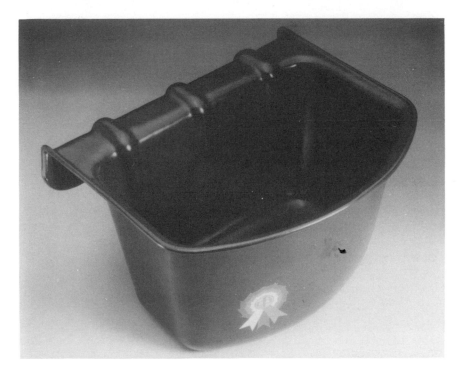

Plate 38 This useful manger can be hooked over a post and rail fence. (*Photograph supplied by Ash Plastic Products.*)

Always feed all horses at the same time to prevent fights, or remove an individual to be fed from the field and feed it out of sight of the others.

Some horse farms, especially in the USA, employ long troughs for feeding. This is not an ideal arrangement as it can lead to bullying and precludes individual rationing. If you *must* use this method, provide ample feeding space per horse.

It has been found with trough feeding that the boldest greedy feeder in the paddock can wind up with carbohydrate overload, through bolting its feed and eating too large a quantity at one time. This can be fatal, or at least lead to colic, muscle myopathies or laminitis (founder). One way to reduce the problem is to add chaff to bulk out the ration. Another suggestion is the use of cereals which have been treated by the extrusion method (plate 39). This flash cooking expands the grain, making it bulkier and necessitating slower eating, and trials in the USA indicate promising results. Plate 40 illustrates the Orbistore feed store, which can be sited out-of-doors.

Plate 39 One of the problems of feeding groups of pastured horses on cereals (e.g. yearlings) is that the boldest may overeat, occasionally with fatal results. The use of extruded cereals, such as this barley, may help to solve the problem as they are eaten more slowly.

Plate 40 The Orbistore feed store is useful as it can be sited out-of-doors, adjacent to paddocks if desired. It is weather- and vermin-proof, and because it is round, stale feed does not tend to accumulate in the corners. (*Photograph supplied by Glasdon Products of Blackpool.*)

Automatic feed dispensers

An automatic feed dispenser which is battery operated and may thus be used in a field shelter has recently been developed in the UK. Called the Equi-Feeder it is designed to disperse three cubed feeds per day at pre-set times, and may thus be useful for the working horse owner. I would suggest it should only be used if there is just one horse to feed as fights could occur, especially if, due to batteries running low or wrong settings, feeds from several machines were dispensed at different times. Plate 41 illustrates a hydroponic unit.

Plate 41 Hydroponic grass units provide a useful supplementary green feed for both grass-kept and stabled horses in the winter and high summer. (*Photograph supplied by Eco-Grass.*)

Bodyweight tapes

There are a number of tapes on the market for estimating a horse's bodyweight, and it is a good idea to do this once a week as a check on the horse's condition, so that grazing restrictions and hay or hard feed allowances may be modified as necessary (plate 42).

The one I currently use is the Equitape, and others in the UK are the Lincoln Equimeasure and Spillers weight tape (which requires reading the weight off a table). In the USA, many feed merchants provide paper tapes (which are subject to stretching if overused) free with feed purchases (e.g. Purina). As they are all based on different formulae, they all give slightly different readings, so always use the same brand of tape or formula.

The tapes are at best an estimate and no substitute for actual weighing on a weighbridge, and tend to be less accurate for ponies and some cobs and brood mares, and inaccurate for foals, and totally useless for donkeys.

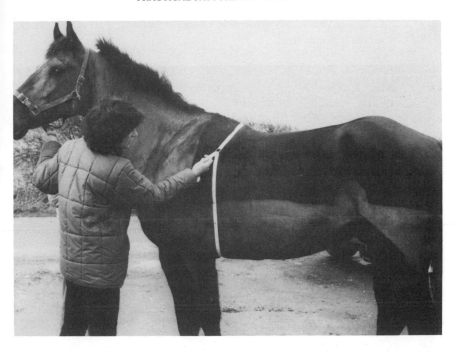

Plate 42 Regular monitoring of bodyweight using a weight tape (e.g. Equitape) will give a guide to any restriction of grazing or additional feeding required, and ensure accurate dosing with wormers.

Chapter 4
The role of pasture
in horse health

As a pasture is a major part of the horse's environment it can exert a considerable influence on the horse's health and well-being, and management of the pasture must be considered as an integral part of management and husbandry of the horse (plate 43). The horse's health, nutritional status and even temperament and mental well-being may reflect the type of pasture it is kept on, and where appropriate these general aspects have been mentioned throughout this book. Plate 44 illustrates the digestive system of the horse.

There are a number of specific conditions and disorders particularly relating to, or which can be influenced by, pasture management, and these are considered in this chapter. The most obvious conditions are those involving parasites.

Parasites

Intestinal parasites

There are a number of ways which intestinal parasites, and their various growth stages, can affect the nutrition and health status of an animal:
(a) Primary damage to the gut wall, reducing absorption of nutrients.
(b) Secondary damage to the gut wall, e.g. by stopping or restricting blood supply to a part of the gut, leading to necrosis (death of tissue), and therefore less efficient feed utilisation, and possibly colic.
(c) By actually causing a blockage at some point in the gut, e.g. a heavy bot burden in the stomach of a foal, again leading to colic.
(d) The parasites' use of nutrients which would otherwise be used by the horse.
(e) They may cause blood loss, externally by blood-sucking parasites, such as lice, ticks, mosquitos, or internally (e.g. redworms), leading to anaemia which can be sufficiently serious to cause death, and will at least lead to unnecessary fatigue.

Plate 43 Neglecting pasture management can equal neglecting the horse. Just because the field is green, it may not provide an adequate environment. (*Photograph supplied by the Horse And Ponies Protection Association (HAPPA).*)

Virtually all equines have some sort of parasite burden, whether or not some form of parasite control is practised. The degree to which this affects the horse's health, longevity and performance is very much in the hands of the horse owner.

A brief understanding of the life cycles of the main internal parasites is an aid to understanding the approach to their control. There are two aspects of parasite control, one is pasture management, and the other is chemical control with anthelmintics in the horse (or donkey).

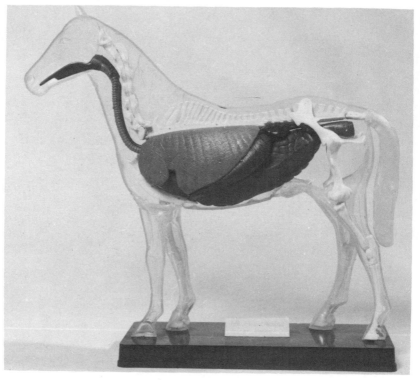

Plate 44 The horse has evolved as a grazing animal and its massive large intestine helps to ferment and digest, with the aid of symbiotic micro-organisms, the fibrous herbage in the diet. (*Photograph taken by Gerald Brownsill.*)

Worms

Large strongyles (large redworms, *Strongylus* and *Triodontophorus spp.*)

These nematodes are the most dangerous parasites of horses, and although there are some fifty species of strongyle (large and small), the most damaging of the ten large strongyle species is probably *Strongylus vulgaris* (also known as *Delafondia vulgaris*).

Large strongyles are reddish-brown worms from 2–5 cm in length. The mature stages are found in the large intestine and caecum (plate 45). Their life cycle is some six to eleven months. The mature large strongyles feed on the intestinal lining. The females deposit eggs which are passed out in the faeces. The eggs may hatch in as little as three days, and grazing horses swallow the infective larvae. The larvae penetrate the gut wall, and with the exception of *Triodontophorus spp.* migrate through the arteries, the liver and other tissues.

Strongylus vulgaris is particularly dangerous because of the tendency for its developing larvae to pass through the mesenteric arterial system

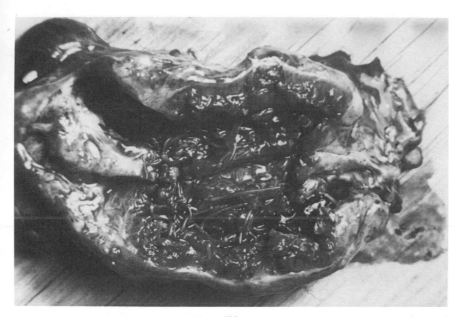

Plate 45 Large redworms in the intestine. (*Photograph supplied by M.S.D.-Agvet.*)

which supplies blood to the digestive tract. There they cause arteritis and thrombosis which may have serious and sometimes fatal consequences, when the thromboses or clots break off and cause blockages in the arteries leading to the intestines and also to the hind legs. Besides irritating and damaging the arterial walls, the artery walls may 'balloon' into an aneurism (collapse or bursting of the artery).

Typical signs of infestation are diarrhoea, fever, oedema, loss of appetite, depression, weight loss, and hindleg lameness. The most characteristic sign is chronic colic, which is not relieved by any of the usual colic treatments, or an acute attack, perhaps triggered by a change in feeding or management. It has been suggested that infestation by large strongyle larvae may be implicated in up to 90% of colic cases.

Small strongyles (small redworms – nematodes)

These are the least pathogenic strongyles, although generally most abundant. There are forty or so species of small strongyles (subfamily *Cyathostoninae*) which undergo larval development in the wall of the caecum and colon.

The worms range from 4–26 mm in length, and their life cycle takes some six to ten weeks. These parasites undergo a typical direct nematode life cycle. The eggs of the mature females are passed out in the horse's faeces. Within a week the infective larvae develop and contaminate grass, bedding, water or feed, whence they are consumed by grazing horses. The larvae then burrow into the gut lining and emerge as egg-laying adults.

Small strongyles generally occur in mixed infections of several species. The damage they cause is largely due to the activity of adult worms in the large intestine, although the larvae may suck blood and induce anaemia. Symptoms of infestation are diarrhoea or constipation, loss of appetite, again colic, and weight loss due to serious impairment of digestive function in the large intestine in heavy infestations. Foals are particularly susceptible to small strongyle burdens.

Pinworm (*Oxyuris equi*)

Horses with chronic infections have a very poor appearance. The male pinworms are 9–12 mm long, the females are up to 150 (egg-filled) mm in length, having a much larger body and pin-like tail. The life cycle lasts some four to five months. Adult female worms migrate out of the anus and lay clusters of sticky eggs on the skin surrounding it. Depending on the ambient temperature, these eggs become infective in three to five days. The horse may increase its infection by biting at the itching larvae on the skin surface, or rubbing them so they drop off and land in feed, bedding (which may be eaten), pasture or water. The larvae thus consumed penetrate the lining of the large intestine to complete their development, whence they emerge into the gut as a new generation of egg-laying adults.

The effects on the horse range from intestinal ulceration, which may be aggravated by stress, especially in foals, to intense itching, loss of condition and poor appearance. Biting, rubbing and scratching may cause wounds which are open to infection, dull coat and loss of hair known as 'rat-tail'.

Roundworm (ascarids – *Parascaris equorum*)

These can have serious effects, particularly on foals aged six months or less. Infestation causes severe debility, poor growth, weight loss and even death.

These heavy-bodied worms may be up to 50 cm (yes, 20 *inches*) long. Their life cycle lasts some ten to twelve weeks, and is again typical of nematode worms. Eggs passed in the faeces are highly resistant to drying out (e.g. by scattering droppings in the field with a harrow). The second-stage larvae develop inside the egg on pasture within six weeks. When swallowed by the grazing horse (or even from hay) the larvae hatch and penetrate the wall of the intestine and are carried by the circulatory (blood) system to the liver, heart and other organs, where they continue to develop. Eventually they migrate to the lungs. Here they break out through the lungs and migrate up the trachea (wind pipe). They are then coughed up and swallowed, maturing in the small intestine into egg-laying adults.

The adult worms can cause bile-duct and intestinal obstructions, and occasional gut perforation. Damage is more widespread in foals. Also, large numbers of larvae may simultaneously break into the lungs causing haemorrhaging and coughing. Acute infection accompanied by severe enteritis may result in alternating diarrhoea and constipation. Heavily infected foals may become debilitated and lose weight. Lung damage in foals causes 'summer colds' with coughing and fever, and loss of appetite, and may be followed by secondary infections such as pneumonia.

Stomach hairworm (*Trichostrongylus axei*)

The stomach hairworm is rarely a problem alone, but contributes to burdens of mixed worm species.

The adult worm is tiny (0.5 cm) and hair-like. The life cycle lasts three weeks and is again typical of nematode worms. (Somebody once said that if you took away everything from the earth other than nematode worms, a ghostly outline of the world, the land, the sea, fish, animals, even men, would be visible as a shimmering mass of worms! Yuk!)

The hairworm larvae develop to the infective stage in four to six days on pasture in warm, humid conditions. The hairworms irritate the gut lining, damaging the capillaries and lymphatic systems and causing blood and fluid loss into the gut. Blood loss can cause anaemia, oedema and rapid loss of condition. Dark-coloured diarrhoea is characteristic and results from the diminished capacity of the damaged intestine to absorb digested feed. Foals are particularly susceptible to infestation.

Large-mouthed stomach worm (*Habronema spp.*)

Larvae in the horse's skin can cause persistent summer sores, both annoying and disfiguring the horse. Stomach infection with *Habronema* can cause gastritis (inflamed stomach) and occasionally ulceration or even rupture of the stomach.

The adult worms are lightish in colour and 10–25 mm in length. They have a secondary host, the housefly or stablefly, which are both found on pastures, which is essential to completion of their life cycle. Thus, control of flies by whatever means, including for stabled horses a feed-through fly repellant (e.g. Equitrol by Farnham, which is fed to the horse but passes straight through into the dung, where it prevents maturation of flies, an

American product not available in the UK) is an important factor in controlling these parasites.

The life cycle consists of ingestion of the larvae by housefly or stablefly maggots feeding in manure. The larvae develop within the maggots and become infective as the adult fly emerges from its pupa. These infective larvae are then deposited on the lips, nostrils and wounds of horses as the flies feed. If they are licked or swallowed by the horse, the worm larvae mature in the stomach to complete their life cycle. The new fly-repellant tags which may be fitted to the horse's bridle, headcollar, hock or tail are therefore a new weapon in the armoury against these troublesome worms, as well as the obvious annoyance and worrying of the flies themselves.

The effect of the large-mouthed stomach worms on the horse are that tumour-like growths may form on the stomach wall, and these occasionally rupture or block the passage of food from the stomach. Also, worm larvae deposited in skin wounds migrate and feed causing persistent summer sores. These infected wounds may heal in the winter but may recur when the warm weather returns. Larvae may also be deposited by flies in the eye, where they cause wart-like lesions, watering eyes and sometimes sensitivity to sunlight (photophobia).

Neck threadworm (microfilariae of *Onchocerca spp.*)

Heavy infestations can be debilitating to the horse and may restrict freedom of movement of the neck or forelegs. Not strictly a parasite of the intestinal tract, but requires similar treatment, and should be looked at in conjunction with sweet itch (below) as the *Culiculoides* midge is implicated.

Tapeworms (belong to the class *Cestoda*)

These worms have flat, segmented bodies, without cavities or an alimentary canal. Two types affect horses – *Anoplocephala* and *Paranoplocephala*. They are distinguished from tapeworms infecting other species because they have no protrudable part (rostellum) nor hooks on their heads (scolex).

Anoplocephala magna is up to 80 cm long and 2 cm wide, and lives in the small intestine. *A. perfoliata* is up to 8 cm long and 3–4 mm wide with a small head, and infests both the small and large intestines. *Paranoplocephala mamillana* is 6–60 mm long and 4–6 mm wide, and lives in the small intestine.

Each type consists of a number of segments containing both male and female reproductive organs. The eggs may be self-fertilized, or cross-fertilized by male gametes from other segments, and stored in the uterus of the segment, which then breaks off and passes out in the horse's faeces. An intermediate host such as the oribatid mite feeds on the eggs, which develop into cysticeroids (bladderworms). When the mite and the cysticeroids it contains is swallowed by the horse, the cysticerus ruptures, releasing an adult tapeworm segment.

Light infestations do not produce symptoms, but cause ulcers in the intestine. Large numbers cause digestive disturbances, a shaggy coat and anaemia, and have been known to rupture the caecum.

Liver fluke (*Fasciola hepatica*)

Rarely significant in horses, although a few individuals appear to be susceptible. The horse is not a natural host for liver fluke.

Lungworm (*Metastrongylidae Dictyocaulis arnfieldi*)

Another nematode of the order strongyloidea, the lungworm is a slender roundworm which lives in the air passages (bronchioles) of the lungs, where it lays eggs. These are coughed up and swallowed. The type affecting horses and donkeys, *D. arnfieldi*, has a direct nematodal life history and does not require an intermediate host.

The adult male worm is about 36 mm long and the female 60 mm. The eggs laid in the lungs are coughed up and swallowed, and pass out intact in the faeces, where they hatch into larvae, which are then grazed by the horse or donkey. These larvae then bore through the intestinal wall to the mesenteric lymph nodes and enter the lymphatic system and on into the veins. Venous blood carries them back to the heart which pumps them back to the lungs. Here they develop into adults which lay eggs, and so forth. They irritate and inflame the bronchi, causing bronchitis, coughing and pneumonia, and the larvae may cause diarrhoea.

Donkeys can carry a heavy burden without showing symptoms, and are often blamed exclusively for infecting horses. However, there are instances of horses (including one very valuable thoroughbred stallion) which can act as carriers without showing symptoms.

It is perfectly possible to eradicate lungworms from donkeys and render them safe for grazing with horses, provided of course a continuous regime of overall parasite control is practised for both the horses and the donkeys.

According to The Donkey Sanctuary at Sidmouth in Devon, between 60 and 70% of all donkeys in the UK, which have not received specialist anti-lungworm treatment, carry lungworm, although they usually exhibit no clinical signs of infestation. Virtually all anthelmintics available in the UK *do not* control lungworm in either the donkey or the horse when used at a normal dose rate, with the *possible* exception of Ivermectin. Some wormers, when used at a standard rate, may produce a very short-lived interruption in the production of young larvae, but within a few days the 'parents' start reproducing again. Most horses (probably 99%) which live apart from donkeys do not have lungworm, and have virtually no resistance when exposed to it.

There are available special wormers (and some standard wormers using special techniques) which can be relied upon to remove the lungworm from both donkeys and horses. However, some of these techniques do not have a normal pharmaceutical licence so the veterinary practitioner concerned would require permission from the owner to use the technique. Treated animals should be isolated from untreated ones, and only allowed to graze on pastures *known* to be free of larvae (see below), and then only when tests have shown the animal concerned to be *completely* free of lungworm infestation.

Pastures grazed by infected animals can carry infection to other equines for many months after the initial culprit has left. However, special management techniques can be employed which will render such pastures clean again (which may take up to two years).

Owners wishing to import a donkey into a horse population should keep it isolated until it has been specifically treated for lungworm, and has been subsequently tested by a reliable laboratory to prove its 'clean' status. Once it has been properly treated, and allowed to graze only clean pastures, it will present no risk to horses.

The Donkey Sanctuary, through its Senior Veterinary Surgeon J.N. Fowler, BVet.Med. MRCVS, is happy to act as a free 'second opinion' service, their advice being channelled through the consulting clinician.

Insects

Warble fly (*Hypoderma spp.*)

Not a gut parasite, although the gut is part of the life cycle and may become damaged. Two species are significant in the UK because their larvae (warbles) occasionally live in horses, although cattle are the natural hosts. *Hypoderma bovis* and *H. lineatum* are found all over Britain, except in Orkney and the Shetland Isles, and are also found in Continental Europe, Canada and the USA.

Warbles are large flies, resembling bees. *Hypoderma lineatum* is black with yellow or orange bands and *H. bovis* is black and yellow. They lay eggs which stick to the base of the hairs of the horse's coat, *H. lineatum* laying eggs in rows of up to twenty on each hair, and *H. bovis* laying eggs singly.

Three to six days later larvae, about 0.5 mm long, emerge from the eggs and develop into maggots. These crawl down the hairs and burrow into the skin, migrating through the horse's body, usually spending the autumn in the tissues of the leg, chest and belly. They then migrate to the oesophagus. In February (UK) they migrate to the skin along the horse's back, causing a swelling with a small hole, through which the maggot breathes.

Each maggot forms a small abscess and one to eight days later moults to become the third and last larval stage. The translucent larvae feed on the pus produced as a response by the horse, and turn dark brown. After about thirty days they leave through holes in the skin, fall to the ground and pupate. The adult flies then emerge in May. Treatment involves skin dressings along the back. *Do not* squeeze the larvae, although warm poultices can be used to encourage them to emerge.

Bots (*Gastrophilus spp.*)

The reddish mature bot fly larvae is 2 cm long with a broad, rounded

Plate 46 Bot larvae in the stomach develop from the yellow eggs seen on the horse's legs and belly in summer, and are not controlled by many broad-spectrum wormers. (*Photograph supplied by M.S.D.-Agvet.*)

body and narrow, hooked 'head'. The larvae overwinter attached to the mucosa of the horse's stomach, and may be so numerous that they completely cover the stomach lining. In the spring, they release their hold and are passed out in the faeces. They then burrow into the ground and pupate. The adult flies emerge in three to ten weeks. The female flies pester horses, laying and firmly attaching eggs singly on to individual hairs, particularly on the legs, belly, chin and throat. These are the little yellow 'dots' that are visible to the naked eye, and may be removed by shaving with a razor. The legs may be kept dressed with liquid paraffin to make the hairs 'slippery' and difficult to attach the eggs to. Do not assume you have removed all the eggs, however.

After hatching, the larvae are licked into the horse's mouth and burrow into the tongue and gums. The discomfort and 'itching' this causes may make the horse or pony 'yawn' or 'laugh' provoking, unfortunately, an 'Oh, isn't he sweet?' response instead of a more desirable 'We must dose him for bots'. After a month, the larvae migrate to the stomach lining.

Migration by larvae through the mucous membranes of tongue, lips and gums causes lesions which are open to infection. Attachment to the gut will cause haemorrhage, ulceration or even perforation of the stomach, particularly in the foal, resulting in fatal peritonitis and intestinal blockage. Also, attempts at egg-laying by adult flies cause nervousness and irritation.

Prevention and treatment of intestinal parasites

This falls into two main categories: (a) managing the horse, and (b) managing the horse's environment – paddocks, feed supply etc., anthelmintics (wormers), fly control and so forth. The degree of parasite burden carried by the horse can be strongly influenced by the way his environment is managed. In the horse, because of the enormous variation in length of life cycle, and the ineffectiveness of many anthelmintics (wormers) against larval stages of worms and flies, dosing must be regular and frequent. A scrupulous plan should be drawn up in conjunction with your vet.

As worms can become resistant to different wormers, you may devise a rotation of different chemicals (not just different brands which may contain the same chemical). There is considerable difference of opinion amongst experts as to whether these changes should be at each worming, every few times, or once a year. Be guided by your veterinarian. One new wormer, Ivermectin (e.g. Eqvalan) to which there is no known resistance can, in theory, be used all the time. However, I have some reservations about this.

Ivermectin is plainly an excellent product, but as it is *so* effective, if used from foalhood, it may prevent the foal from developing resistance to a parasite challenge (as occurs in older horses to some extent). Now this won't matter if the foal is *always* going to be treated with Ivermectin, but I am concerned that, say the foal is kept and regularly wormed with Ivermectin until it is four years old, and then sold to an owner who doesn't worm his horses, or not very often or effectively. The animal may then be particularly susceptible to attack by intestinal parasites.

If you *do* use Ivermectin all the time, you could probably reduce doses from four to every six or eight weeks, depending on the animal's environment, because it kills the *larval* forms of most of the parasites mentioned above at *normal dosage rates*. Otherwise, for performance horses, foals, brood mares (check for contra-indications), and animals on horse-sick pastures (e.g. many riding schools and livery farms), worm every four weeks. Other horses and ponies may be wormed every six to eight weeks, but remember there may always be larval stages developing which the wormer will not kill (unless it is Ivermectin).

If you are presented with a very badly infested horse, it may be prudent not to worm with Ivermectin the first time 'to knock everything out' as there would be a lot of disintegrating larvae floating about which *could* lead to blockages or oedema. To be on the safe side, I would use an anthelmintic which would knock out the mature worms, and follow up five to ten days later with either Ivermectin or a high dose (strictly under veterinary guidance) of another anthelmintic to eradicate the larvae. Then go on a proper, four-weekly regime to prevent re-infestation.

If a horse has significant damage to the gut it may suffer from malabsorption syndrome, and will require a nutritious, easily digestible diet, possibly with additional protein over and above the normal requirement for mature horses. I expect part of the reason for loss of condition in elderly horses may be due to a lifetime of gut damage, or even damage incurred as a foal.

It is a waste of time feeding any ration, however good, to a horse, pony or donkey, stabled or at pasture, if you do not practise a comprehensive parasite control programme, which should be devised in conjunction with your veterinary practitioner, as many over-the-counter preparations are *not* fully effective against all species, or larval stages.

There is little to be gained from faecal worm counts, as it is in the nature of things that one day there may be virtually none, and the next, thousands. You might just as well spend the money on a comprehensive anthelmintic and have done with it. Blood tests may also be done for redworms, but again, they cost about the same, or even more, than an anthelmintic, so are frequently rather a waste of time.

Plate 47 Regular dosing with anthelmintics (wormers) is essential for management of horses at pasture. This multi-dose syringe is suitable for owners of a number of horses. (*Photograph supplied by the Wellcome Foundation.*)

As far as the horse's environment is concerned, much can be done to reduce the challenge from parasites. For horses at grass, the best method is to collect *all* the droppings, twice a day, by good old spade and wheelbarrow, or purpose-designed 'vacuum cleaner'. Less frequent collection may be more convenient, but is considerably less effective. *Do not* use stable manure to fertilize horse paddocks. Even if it is extremely well-rotted it may contain encysted eggs or larvae. Try and swap it with Farmer Bloggs for some cattle manure!

The next method is a contentious one. You can harrow, daily or weekly, your paddocks to spread out and dessicate the droppings and their eggs. This is *only* effective if done in hot, breezy, drying weather, and some people consider that all it does is spread out the eggs for maximum infestation, as horses do not normally graze where they have dunged, so spreading the droppings forces them to consume grass that has had dung spread on it.

I have an open mind about this, and still tend to prefer harrowing in suitable weather, as it also helps to aerate the grass. I suspect that if you harrow *and* graze with cattle, this will be the most beneficial, as cattle are not host to most horse parasites, so when cattle eat parasite eggs/larvae, they break their life cycles and kill them, as to a lesser extent, do sheep.

If you are not using cattle, then *only* harrow if drying conditions prevail. Using cattle without harrowing will also help, although it is not wise to graze cattle with valuable bloodstock or young animals. It *is* good management to let them follow horses in a rotational grazing system and 'mop up' the paddocks.

Obviously if horses are kept on heavily-stocked or horse-sick paddocks or are constantly exposed to new horses (e.g. dealer's yards) they are likely to be re-infested more often. As we have seen, some parasites include flies of various types in their life cycle, so prevention of flies in buildings, and on the grazing horse, is of great importance. All horses at pasture should be wormed every four to eight weeks, depending on stocking rate.

The Donkey Sanctuary offers the following advice for eradication of lungworm (*Dictyocaulus arnfieldi*) in pastures:

'The first step in the management of a field known to be, or suspected of being, contaminated by equine lungworm, is to remove all equines. Such animals should be housed, treated and kept housed until proven to be clear of infection, and then allowed to graze only clean fields.'

Infested fields can be cleaned of lungworm using combinations of the following methods:

(a) Intensive grazing with sheep or cattle. Equine lungworm does not affect these species. Sufficient sheep (or cattle) should be employed to eat the grass down really tight, including rubbishy grass in the base of the hedges and fences. If the sheep are temporarily removed, the grass should not be allowed to regrow to more than three inches before the sheep are returned.

(b) The pastures should be harrowed as often as possible (perhaps once a month) to elevate the dead 'mat'.

(c) Many lungworm larvae will not survive winter conditions. The 'clean up' programme must be allowed to continue over at least two winters without grazing equines.

All dung heaps and muck heaps in the field should be removed. Exploding dung (!) fungus seed pods will help spread lungworm larvae.

Or

(a) As an alternative to grazing sheep, the fields could be sprayed with a total herbicide, and then allowed to remain fallow for at least six months including the course of one winter. Particular attention must again be paid to hedge bottoms, fences and gateways.

(b) After reseeding, such pastures should again be grazed by sheep, or hay cropped. Any hay made from fields during the initial two years of the clean-up programme should be fed to cattle or sheep only.

No fields should have equines grazing on them during the initial two years of the clean-up programme, whichever methods are used. If ploughing is employed the fields should be kept to an arable regime for a three year period. The subsequent initial grass or hay crop should not be used for equines. Where it is impossible to remove all equines, the pastures should be divided into two, with a 4 m (12 feet) 'No Man's Land' down the dividing line. One half can then be cleaned up over the initial two years. (The 'No Man's Land' should either be sprayed, or kept permanently tilled.) All involved equines should be retreated after two years and then transferred to the clean half. The procedure can then be repeated for the second half.

Giving wormers (anthelmintics)

There are various forms of anthelmintic, most of them unpalatable. They may be given as powders or pellets on the feed, which means they may be left or even sneezed away; by injection, which can lead to a skin reaction and requires scrupulous attention to hygiene (particularly use of a new needle for each horse, as fatalities have occurred where this has not been practised), injections are most useful for unhandled animals using a crush; by stomach tube – a traumatic method rarely justified with the advent of paste wormers; by paste applied via a graduated syringe to the back of the tongue; by self-ingestion using a feed block containing a wormer – not yet developed for horses, but Crystallyx blocks for cattle (made by Pfizer), containing Panacur, are a potentially useful method for range animals.

The paste wormers are usually the most convenient. You just squirt the (usually reasonably palatable) paste onto the back of the tongue, so the horse can't spit it out. If the horse has not been handled sufficiently, a friend has devised a new technique. She squirts the paste on to brown bread which has been 'buttered' with minty toothpaste, and either feeds it by hand as a treat (sneaky, huh?) or chops it up and feeds it with cubes.

If you are using a granule, such as the boticide dichlorvos, which gives off a noxious gas, chill the wormer overnight in the fridge, and that will reduce the taste and smell, without significantly affecting its efficacy. Also, when giving wormers in the feed, you can disguise them with molasses, fruit juice, Ribena (blackcurrant squash), honey or spices, but make sure the horse likes the taste first or you may waste an expensive worm dose.

Always follow the instructions of your veterinarian and/or pharmacist, and follow the directions on the pack. Do not give other species (e.g. dogs) access to the product as they may be allergic. Dose according to bodyweight.

Diseases involving external parasites affecting pastured horses

Sweet itch (sweat itch, summer itch (Europe), Queensland itch (Australia))

This is *not* caused by feeding, although many horse owners think it is. It is in fact an allergic reaction, possibly complicated by hypersensitivity to sunlight, to the bites of midges of the *culiculoides* species (in the UK, *Culiculoides pulicaris*).

Sweet itch is a type of dermatitis, usually confined to the mane and tail, but which may spread along the back. The skin becomes thickened and scaly and may have small pustules on the surface. The condition is intensely sore and irritating and the hairs of the mane and tail break off, or are rubbed off at the skin surface, giving a 'rat-tailed' appearance. Rubbing may lead to the development of large sores.

Various soothing lotions and ointments have been produced, to be combined with cortisone and antihistamine injections, and a vaccine is being developed. It has also been suggested that feeding quantities of garlic may have the effect of deterring the midges, and a number of my clients have tried this, with some success.

Judicious stabling of animals and use of fly-repellent strips in the

Plate 48 Many horse owners have found that feeding garlic helps repel flies! Garlic is also thought to have other beneficial effects on health.

stable, when the midges are most active, i.e. May and September (UK) at least from 4 p.m. until dark, will help reduce the problem. Liquid paraffin may be applied to the coat to make it harder for the midges to crawl to the skin to bite – it is unsightly, but so are sweet-itch sores. Fly repellents tend not to be very long-lasting, although the new ones on impregnated strips which can be attached to the head collar, hock or tail may well be a more effective deterrent. Check with your vet before using any fly repellents on open sores.

The ability to develop this allergy is inherited so it is irresponsible to breed from affected animals.

Mud fever and rain scald

Mud fever and rain scald are both dermatological infections which occur particularly in pastured horses; the former mostly in winter, the latter in warm, humid summer conditions. Rain scald can occur if ponies spend hours standing under dripping trees.

Mud fever ('grease') and rain scald are two distinct conditions, but both are caused by the organism *Dermatophilus congolensis. Dermatophilus* attacks the surface layer of the skin, leaving it raw and oozing. Predisposing conditions for mud fever are horses constantly standing in wet, muddy conditions, leading to cracked skin over the heels which may help the organism to 'gain a foothold'. It has been suggested that the organism does not occur in some soils and there are certainly some areas where it seems to be less prevalent than others. However, it certainly makes sense to keep horse paddocks well-drained and avoid muddy conditions, and clean and dry the heels when horses are brought to stand in a stable, protecting the heels with Vaseline when they are turned out again. Mud fever is most common in winter and is sometimes misdiagnosed when in fact the problem is photosensitisation or an allergy to sugar beet.

Rain scald occurs over the horse's back, especially in damp conditions when there is a combination of high temperature and high humidity. 'The infection causes small areas of skin to ooze, and the resulting fluid dries and mats the hairs together, resulting in a characteristic "paint brush" effect. As the rain and sweat run over the horse's body the infection is spread, so that the infection is worst over places such as the shoulders and rump where the rain "runs off".' (From *Horse Ailments and Health Care*, by Colin Vogel.)

Mud fever is not as temperature dependent as rain scald and in some countries is more prevalent in spring and summer when the horses are standing for long periods in lush pasture wet with dew or rain. The resultant cracks in bad cases can open up an entry for serious, *even fatal* infections if septicaemia ('blood poisoning') sets in, and it is *always* wise to call a vet.

Treatment includes moving to a drier location, drying legs, clipping off all the hair in the area and removing the scabs. The reason for the latter is that the scabs protect the area of skin where the bacteria live. If the scabs are removed the area dries out and the bacteria can no longer survive. This may need to be done several times a day at first, and the area will look inflamed and worse afterwards, but don't let this put you off. Once this has been done, a mild antiseptic, such as 1% potash alum or 0.5% copper sulphate solution, will kill off the remaining bacteria. It is a waste of money to apply expensive creams if you do not also remove the scabs. Leaving just one scab can result in re-infection. Use a barrier cream to exclude moisture once the horse is cured to prevent a recurrence. If the case is serious, your vet will prescribe antibiotics.

Ringworm

This is a fungal infection caused by several different fungi. Once the fungus becomes established on the skin surface, it releases spores which spread the infection. The name ringworm derives from the fact that the affected hairs break off at their bases, leaving a crusty bald patch of skin which is *usually* round. It can affect any part of the body.

Cattle and humans may also be infected (the former often being the initial source of infection) and infective spores may remain on fences, brushes, rugs and tack for many months, so paddocks and equipment which have been used for infected animals should not be used for non-infected animals and should, where practicable, be cleaned.

In the UK, the antibiotic Griseofulvin may be fed for seven days to clear up infection, and topical application of tincture of iodine or gentian violet is also popular. In the USA, Captan is used as a topical treatment for both livestock and fences. Of other topical treatments, only the antibiotic Natamycin is generally considered effective.

Always check any new stock before introducing them into your paddocks and isolate and treat infected animals.

Flies and insects

Apart from specific parasites mentioned above, constant irritation of pastured animals by flies can lead to considerable loss of condition and even exhaustion as they stamp and gallop about to avoid them, and can reduce grazing time. Animals pastured alone are particularly at risk as they cannot stand head-to-tail with a companion for mutual fly swatting.

Picking up droppings will help to reduce the fly population to a degree, but it makes sense to give pastured horses some sort of protection against flies and, when they are at their worst, even to bring them in in the day time. Otherwise, there are numerous fly sprays and wipes on the market, and long-acting gels which are applied along the spine every few days and gradually work over the whole coat.

Other useful ideas are fringed brow bands on headcollars, insecticide-impregnated brow bands, hock bands and tail tags, which are especially useful as the fly repellent is deposited on the flanks, etc., when the horse swishes its tail. There are also net face covers which can be attached to the headcollar, with or without ear covers. Although I have found these very useful out riding, I am rather concerned about the safety of turning a horse out in one, especially with ear pieces, as a horse could be driven demented if flies crawled inside the net, and up round the ears, and couldn't get out. I have never actually heard of this happening, but would suggest it is a very real possibility. However, I certainly found an 'ear-less' one *very* beneficial on a normally head-shy horse, who accepted it straight away and seemed visibly relieved, for riding out.

Fly repellents vary considerably in their efficacy and durability and it

Plate 49 An 'anti-fly' net can bring welcome relief to grazing horses when attached to a head-collar. Choose the type without ear-pieces so that flies have an 'escape route' if they crawl underneath. A fringed or impregnated brow-band might be safer for unsupervised horses.

is worth experimenting to find the one that suits you and your horse. Aerosols should be avoided as they can frighten a horse and are thought to be damaging to the environment (by reducing the ozone layer which protects us from the strongest rays of the sun). Squeeze bottles may also frighten some horses, and sprays should never be used directly on the face.

Impregnated fly wipes are very useful as they can also be tucked under the brow band, and there are various gels and liquids, including oil of citronilla, which can be applied with a cloth or sponge.

Be especially careful to protect around the eyes, without getting chemicals *in* the eyes, as flies can set up a nasty and potentially damaging eye infection.

Poisoning

Poisons may be defined as substances which harm the body externally or internally and which act by interfering with normal metabolic processes in the body. Poisoning may be temporary, but may result in permanent damage, particularly to the liver, and even death.

Whenever poisoning is suspected, the vet should be called, and an attempt made to establish the cause prior to his arrival (check for lead paint on fences, poisonous plants, access to pesticides, drugs, e.g. growth promoters in cattle or pig feeds such as monensin sodium (Romensin, Rumensin) or lyncomycin, mouldy feeds, etc.) so he has some idea of what he is dealing with. Comprehensive tables of poisonous plants which may be encountered may be found in chapter 2 of this book.

Substances may be harmless in some circumstances, but highly toxic in others. The majority of plant poisons are some form of toxic alkaloid or glycoside, or other phytotoxins.

The other common form of poisoning is due to ingestion of inorganic materials including heavy metals such as lead, arsenic, or cadmium, e.g. when human sewage has been used as a fertilizer. Inadvertent ingestion of agrochemicals is also likely to cause poisoning. Ask neighbouring farmers to warn you when they are spraying their fields, and beware of spray drift.

Some fertilizers may also cause poisoning, for example nitrate poisoning, the results of which may be immediate, or *may* manifest themselves as unthriftiness or possible abortion. *Do not* fertilize horse paddocks as if they were for dairy cows, and both when using agrochemicals or fertilizers, ensure that horses are kept off the paddocks for at least the minimum period specified by the manufacturer (three to six weeks, preferably including some good rains to break down materials, wash them into the soil and encourage uptake by plants).

Apart from substances which burn (strong acids and alkalis), most poisons must be absorbed into blood to be toxic, having first entered the body by being eaten/drunk, inhaled, absorbed through the skin, or injected. As it is the liver which detoxifies, or tries to, such materials, it is usually damaged or, at the least, inflamed. Some reactions are complicated by the effects of sunlight (photosensitivity), e.g. some anthelmintics, or leguminous phytotoxins.

Symptoms range through diarrhoea, convulsions, coma, muscular inco-ordination, dilation of pupils, constriction of pupils, distressed breathing, sensitisation, and blood in urine. *Always* call the vet. Mycotoxicosis has been discussed elsewhere in this book.

Grass sickness

Grass sickness is a usually fatal, major disturbance of the alimentary canal (gut). It was first reported in Scotland in 1907 and although it is most common in Scotland, it also manifests itself in Southern England (increasingly so), Northern France, Ireland and possibly the USA.

The cause has not been established, although viruses have long been suspected. It almost exclusively affects grass-kept horses, sometimes a group, sometimes an individual. Whether or not it is caused by a virus, I suspect that trace element disturbance in the soil, or luxury uptake by certain selectively-grazed plants, or even mycotoxins, may be a predisposing factor.

I have noted that a number of cases have tended to occur either in fields where deep excavations have occurred (gas pipe laying, removal of old buildings, etc.) which can lead to heavy metals being brought up, or general soil and plant trace element disturbances. This may also apply to paddocks downhill from building sites and road construction sites, where groundwater may carry excavated heavy metals to horses' paddocks, which may then be taken up by certain plants, for which individual horses may have a taste. This is purely conjecture on my part, but is certainly not beyond the realms of possibility.

There are four main types of grass sickness:

(a) Per-acute, symptomised by depression, fast pulse, absence of gut movement, vomiting of green, evil-smelling fluid through nostrils, profuse patchy sweating, and usually pain. Most horses die within twenty-four hours.

(b) Acute – symptoms as above, plus muscle tremor and jaundice. Hard faeces can be felt in the colon by rectal examination. Death is likely to occur within twenty-four to forty-eight hours.

(c) Subacute – symptoms as for (a) and (b), although jaundice may be more acute, the horse may stagger, and muscle tremors are particularly notable between hip and stifle, and near the elbow. The appetite is variable, but often non-existent. (Haematocrit rises progressively from 40% to 60 or 70%). Death may occur between ten and twenty-one days.

(d) Chronic – symptoms include firm faeces turning to diarrhoea. Death may occur from exhaustion after twenty-one days. Cases in this latter group sometimes live if nourished via stomach tube, but are unlikely to be capable of hard work again.

The horse should be isolated as the mode of spread of the disease is unknown, and it may be given electrolyte therapy. Some fluid may be siphoned from the stomach to ease the pressure. In groups (a), (b) and (c), once diagnosis has been confirmed, it is almost certainly the kindest thing to euthanise the animal. For animals in group (d) this is probably also the case, unless some role can be envisaged for what will almost certainly be an invalid, should it survive.

Whatever the cause may be, the effect seems to be damage to the nervous system supplying the gut, and symptoms relate to the consequent cease of gut movements. Whether this damage to the nerves is a primary effect or a secondary one has not yet been elucidated.

Digestive disorders

Scouring (diarrhoea)

Scouring may occur a few days after foaling, probably due to an *E. coli* infection in the gut. The vet may give antibiotics and kaolin, or better still sodium montmerrillonite, to try and stabilise the situation, although the prognosis is not always good.

In cases of non-infectious scouring I have found useful feeding both the mare and foal a probiotic supplement (e.g. Transvite: mare 30 g/day for ten days in feed, foal 10 g/day), or, perhaps most successfully, weaning the foal early (e.g. at three months) on to hard feed, or even younger onto creep feed and a proprietary mare's milk replacer with live yoghurt mixed into it. Care should be taken when choosing a probiotic as fatalities have occured when unsuitable types have been used. Those based on *Streptococcus faecium M74* (e.g. Transvite) seem to be more stable than those based on *Lactobacillus acidophilus*. If using an imported product, be aware that those products transported by ship (e.g. across the Atlantic) may well have lost viability and not work, compared for example to those products flown in a matter of hours under cool conditions from Scandinavia (to the UK). This is probably one reason why some scientific trials find no response to probiotics and others get excellent results!

Probiotics can be given mixed with molasses via a 'wormer' syringe. Repeat this as necessary, and if it is suspected that the milk is too rich, either adjusting the mare's feed, putting her on less lush pasture (beware of putting fibre levels up and excessive increase of fat (butterfat) levels in the milk) or hand-milking her and 'watering' the milk before feeding to the foal may also help. Probiotics are 'good bugs' which help replace 'bad bugs' by stopping them from multiplying, unlike antibiotics which kill both types, thus depriving the animal of the benefits of the 'good bugs'.

In adults, severe scouring may be indicative of a number of disorders, but principally enteritis, i.e. inflammation of the bowel or intestinal lining, caused by bacterial upsets, diet, (particularly mouldy feed containing mycotoxins), chemical poisoning, (e.g. chewing lead paint), or vegetable poisoning (poisonous plants).

It is *vital* that you send for the vet at once; do-it-yourself treatments which may not work may mean the vet arrives too late to help. However, for horses that continually scour mildly, e.g. when at grass, try reducing the quality of the grass offered, or feeding some oat-straw at the same time, and again consider sodium montmerrillonite or probiotics.

I would always offer some form of dry forage (hay/oat-straw/chaff) to horses turned out on lush spring grazing. If they are reluctant to eat it because the grass is so palatable, try using tasty molassed chaff. It is also prudent when turning horses that have been stabled over winter and are to be roughed off or turned out for the summer, to do so gradually increasing the time out by an hour or so a day, as this is a *very* sudden change in feeding. Worm with Ivermectin or another larvicide a few days before turning out to avoid increasing pasture infestation.

Colic

There are many diverse causes of colic (belly-ache), but the two most often associated with pastured horses, which can to a degree be influenced by pasture management, are *sand colic* and *verminous colic*. Others include *impacted colic* due to chewing rubberised webbing fences, which may take years to manifest itself, and blockages due to *phytobezoars* and *enteroliths*.

Sand colic

This is in fact a form of impacted colic and occurs in horses drinking from a natural water source with a sandy bottom, or from grazing on sandy soil, or being fed on the ground in a sand arena. The sand collects in the large intestine, in particular the colon; 27 kg of sand were reportedly found in the colon of one (fatal) case.

Care should be taken if building materials, including sand and gravel, are to be stored in a paddock, as horses may ingest considerable quantities of these if preventative measures are not taken. Sandy water sources can be fenced off (see plate 37), but for horses grazing sandy soil, it may be that the best thing is to ensure that the fibre levels in the diet are high to help 'push' the sand on through the gut. This *may* be the one instance where wheat bran (suitably supplemented with calcium) could be desirable, although this is a hotly disputed point amongst veterinary practitioners. (Abnormal tooth wear may also occur in horses grazing on sandy soil for prolonged periods.)

Verminous colic

This is by far the commonest cause of colic, either due to impaction due to sheer weight of numbers of mature worms in the gut, or bot larvae in the stomach, or, alternatively or in addition, blood clots in the arteries supplying the small and large intestine (mesenteric arteries) so that the blood supply to that part of the gut is cut off. The thrombus (clot) usually occurs because the blood vessel is damaged by migrating red worm larvae (see below). The wall of the artery may collapse (an aneurism) which sets up inflammation of the artery (arteritis) causing the blood to clot and adhere to the lining of the artery (forming a thrombus), or it may break up and be carried in pieces (embolism) through the blood stream until they lodge in a vessel, thereby blocking the blood flow to that part of the gut, leading to chronic necrosis (death of cells or groups of cells) of that part of the gut, or acute peritonitis. (Peritonitis is inflammation of the peritoneum – the smooth membrane lining the abdomen, and its contents (i.e. outside the gut walls) enabling one part of the gut to slide smoothly over another) – possibly causing adhesions, whereby part of the intestines stick together or to the gut wall. Peritonitis may also occur due to infection by bacteria, migrating red worm larvae or foreign bodies penetrating the gut wall. It is characterised by grunting, looking round at the flanks, initially raised white blood cell (leucocyte) count, fever, and reluctance to move.)

Phytobezoars

Some blockages have been reported due to 'phytobezoars' up to 5 cm in diameter. What happens is a foreign body, such as a wood chip, may be accidentally or intentionally ingested, and as feed (plant) material accumulates around it, it 'grows' and may eventually cause a blockage.

This is most likely to occur when wood chewing is a problem. The impaction usually occurs when the phytobezoars pass from the stomach into the small intestine.

Enteroliths

These are 'stones' in the intestines and are apparently relatively common in some parts of the world. There is usually a core (e.g. a nail or other hard material at the centre) and then a deposition of minerals such as magnesium ammonium phosphate in layers to form a regular or irregular ball. These *may* be passed when they are small, but if they continue to grow (up to 8–10 kg enteroliths have been reported!) they can cause impaction colic - particularly as they pass from the large to the small colon in the hind gut. Obviously acute obstructive colic is likely to occur in this event. Up to a hundred enteroliths, presumably small ones, have been reported in one horse, and large ones may lead to death due to a ruptured colon.

Apart from the ingestion of the foreign bodies at the centre of the enteroliths, it seems likely that the mineral nutrition of the horse in question may well be implicated, and the mineral levels in both the diet *and* the drinking water should be carefully monitored. It has been suggested that excess levels of magnesium and phosphorus in the diet, e.g. on a diet containing significant quantities of wheat bran, may well be implicated. Some workers have also suggested that an abnormally high iron intake (*not* the nails in the centre!) may also be implicated. One report has suggested that certain breeds (Arabs and part-Arabs) may be more susceptible to this problem, but did not suggest why.

Treatment of colic

The first step is to identify which type of colic is occurring. This is a job for the veterinary practitioner, and it is prudent to call him whenever colic is suspected. *Do not* use proprietary drenches – they may be contra-indicated for the type of colic and make things worse, and there is always the danger of the fluid going down the wind-pipe and choking the horse.

Whilst waiting for the vet, try to prevent the horse from rolling by gently walking it, but be quite clear that walking the horse for hours without treatment is *cruel* and will simply tire him out and lessen his chances of recovery.

The signs of colic are progressive, and are all indicative of pain, and the first priority for the vet will be to control the pain. The attack may start mildly with indigestion and refusal to eat, the horse looking round at its belly and being restless. It may yawn, a common equine response to discomfort, progressing to pawing and bed scraping, kicking at the belly, box walking, griping, getting up and down, violent rolling, lying on its back with patchy sweat showing along the flanks, and a generally anxious demeanour. Heart rate and respiration rate will almost always be elevated and the horse may be visibly panting. If pain is severe the temperature may rise. Gut sounds (borborygmi) may be absent (impaction) or increased (e.g. spasmodic).

Treatment *must* be preceded by diagnosis from the vet and will include pain control in all cases, the pain being caused by gut wall distension by (a) gas or food, causing a stretched peritoneum (in which there are many pain receptor cells), (b) peritonitis, (c) spasm of muscle in the gut wall. In impacted colic the horse may show signs of having a full bladder and straddle in a urinating position, and liquid paraffin and purgative drugs may be given by stomach tube. In tympanitic colic the horse may appear to crouch or straddle its limbs, and may be treated with anti-spasmodic drugs (e.g. Buscopan) by injection or stomach tube.

Most colicky horses recover, although they may be subject to repeat attacks if the poor management which led to the problem in the first place is not corrected. However, the prognosis is bad if pain fails to respond to treatment, if the pulse is fast and weak, if the haematocrit (packed cell volume or PCV) rises above 50% and the mucous membranes of the mouth and eye turn red or purple.

At least 90% of colic cases are a direct or indirect result of poor or unbalanced horse management and are entirely preventable.

Laminitis (founder)

Laminitis is a disease which is widely misunderstood, by both horse owners and many practising vets, and in its commonest form is a classic example of what can go wrong if the digestive micro-organisms are upset.

Laminitis is concurrent with lactic acid build-up in the blood, originating from the gut. It is not always feed-related and may occur after foaling, due to toxicosis from a retained cleansing (afterbirth), inappropriate use of certain drugs and as a sequel to various forms of illness.

Here I am concerned with feed-related laminitis. Most horse owners and some ill-informed vets insist it is due to 'too much protein', probably because of its incidence in ponies on rich spring grass which is high in protein. Lush spring grass is also rich in soluble carbohydrate, and a sudden influx of carbohydrate from whatever source (e.g. breaking into the grain bins in a feed store) has the effect of 'insulting' the gut micro-organisms, unbalancing them and leading to the production of large amounts of endotoxins, including lactic acid. For some reason, this manifests itself by causing inflammation of the fleshy laminae in the feet, leading to heat and considerable pain, because the fleshy laminae are interspersed with hard laminae and the inflammation has nowhere to go. Both forefeet are usually affected, and occasionally the hindfeet, as well.

The attack at this stage is acute, accompanied by a raised temperature, and requires immediate veterinary intervention if serious structural damage to the feet is to be avoided. Hosing or tubbing the feet, or better still, standing in a whirlpool/jacuzzi boot or a cool-running stream will give some relief, and alternating hot and cold packs can help stimulate the circulation in the foot.

Laminitis may be acute or chronic. It may be an isolated occurrence, but some animals, classically, small, fat, 'gets-fat-on-fresh-air' type ponies, are particularly prone to recurrent or chronic attack. The affected animal will be reluctant to move, and when it does, it will walk on its heels. The classic stance is of the pony with its forefeet stuck out in front, trying to keep its weight on its hindfeet. The foot may be warm or even hot and the animal unwilling to lift a foot up and put more weight on the others. Alternatively the horse may 'paddle', shifting weight constantly from one foot to the other. Pulse, respiration and temperature will rise.

As the attack progresses, the white line may separate from the sole, 'seedy toe' may appear (a crumbling of the toe), laminitic rings appear on the hoof, and the hoof wall may even become completely detached, which can be a blessing in disguise as it affords considerable relief from pain. The pedal bone may rotate and break through the sole at which stage the horse is said to have 'foundered'. This can be discouraged by immediate fitting of heart-bar shoes as the attack commences, which transfer pressure from the wall to the frog and reduce stress on the weakened areas of the foot, or in less severe cases by applying wadding to the rear two-thirds of the frog (*not* the sole) and fixing it with a foot-bandage of sticky tape. This avoids the concussion of shoeing an already stressed hoof. Correct fitting of a heart-bar shoe *can only* be done if the foot is properly X-rayed and the toe opened out to allow for drainage. Bedding on deep sand will also gently support the frog.

As hoof-rings appear during chronic cases they will be wider at the front, and the toe will progressively 'curl up' and require careful and judicious trimming.

As circulation is restricted due to inflammation, forcing the horse to walk is often advocated, a case of being 'cruel to be kind', but if possible, I believe it both kinder and more effective to use modern physiotherapy techniques to reduce inflammation (in conjunction with suitable drug therapy) and stimulate the circulation. There is nothing to be gained from forcing a horse to walk once white-line separation has commenced, and it may even do more harm than good.

Animals with small boxy feet (including donkeys) are most at risk.

Care should be taken to make *all* changes in feeding gradual, including introduction to lush pasture. Always feed some hay/straw when doing the latter, and limit access if possible and try to, say, turn out for two hours twice a day rather than four hours once a day. If cereals and other digestible carbohydrate-rich feeds are to be increased, do so very gradually, ensure adequate fibre (hay/chaff) in the diet and, in susceptible animals, consider using a probiotic during the change-over to help stabilise the gut micro-organisms. It has also been suggested that feeding garlic may improve circulation in the feet by making the blood more 'slippery'!

Once an animal has had laminitis, it seems to have a predisposition to further attacks. Keep the fibre levels in the diet high and consider using fat or oil as an energy source. Also, a predisposition to laminitis may be inherited, in that whether or not a horse is a 'good doer' may be partially inherited, as may inability to balance feed with work on the part of the (thick or idle) horse owner.

Now it may be that some 'gets-fat-on-fresh-air' ponies may suffer from a hormonal imbalance, and particularly hypothyroidism and consequent insulin insufficiency. Preliminary studies on the effects of giving thyroxine or insulin tablets to this type of pony look promising. If it is a hormonal problem, then the 'starvation' technique of 'prevention' is likely to be contra-indicated, as it may further push hormone levels 'out of kilter', for example as the pony becomes deficient in certain nutrients, including iodine, essential for efficient thyroid gland function.

Another deficiency and consequent imbalance which *may* be implicated in chronic laminitis is a form of hyperparathyroidism due to calcium deficiency. As a combination of starvation and bran mashes is a common system of 'management' or 'mismanagement' of such ponies, the consequent calcium:phosphorus imbalance may well ensure an acute case will become a chronic and recurring one.

I suggest instead that you feed hay or hay plus oat-straw and a vitamin/mineral supplement. A convenient one is the Salvana Briquette – a sort of horse biscuit which can be fed on its own. If you decide to turn the pony out on a 'starvation paddock', ensure there are no poisonous weeds that it may eat in desperation, provide some low nutritional quality hay or oat-straw (*not* mouldy or dusty), and a suitable feed block/range block (e.g. in the UK, Keep or Derby blocks) to ensure adequate trace nutrient and protein intake. *Do not* use blocks without hay as the pony may gorge on them.

If the pony is to be worked, I have found that for some reason a particularly safe ration may be made up of chaff plus soaked sugar beet pulp and breadmeal (Bailey's no. 1), formulated according to the pony's work requirements. When deciding on the appropriate level of DE, keep on the safe side and aim for 10–20% less (introduced gradually) than you think you need. I have tried this on many laminitis-prone ponies in hard work, such as competition carriage driving, to good effect. I do not know why this diet seems to work, it *may* be something to do with the main carbohydrate being sugar from the sugar beet – with highly digestible fibre – and the breadmeal being a cooked cereal. That remains to be seen.

It is to be hoped that the awareness of correct diet/exercise management, and possible hormone therapy where indicated, will reduce the incidence and severity of laminitis, as our understanding of this common and distressing disorder increases.

The use of anti-inflammatories such as aspirin and Bute at an early stage can do much to prevent permanent hoof damage through founder, and there are vaccines currently being developed in the USA which could be a major factor in preventing this (and other endotoxic) troublesome disease in susceptible animals. However, this is a 'management disease' and drugs are no substitute for prevention through good management.

Nutritional imbalances

Soil type, herbage species and fertilizer usage can all influence the balance of nutrients which a grazing animal receives. Deficiencies or imbalances may lead to general unthriftiness, infertility, *agalactia* (lack of milk), muscle myopathies (e.g. due to selenium deficiency), skin disorders and growth and bone defects (e.g. *hyperflexion* – contracted tendons, *epiphysitis* – calcium, phosphate and copper imbalances, *nutritional secondary hyperparathyroidism* – e.g. through oxalate rich grass species), even anaemia – so the whole horse may be affected.

Heavily-fed mares on good pasture should be watched for *Eclampsia* (similar to milk fever in cows) due to sudden withdrawal of calcium from the blood at foaling. Ensure that calcium levels in the diet of pre- and post-parturient mares are adequate (but not excessive), and for mares prone to this condition consider feeding some of this calcium in chelated form for the first fourteen days after foaling. Check the calcium:phosphorus ratio and magnesium levels in the diet. The soil in some areas is particularly deficient in magnesium, as any dairy farmer will tell you (dairy cows are prone to 'grass staggers' due to hypomagnesaemia), although horses seem to be able to tolerate particularly low levels of magnesium far better than dairy cows.

It would be impossible to give detailed descriptions of the symptoms of all such conditions in this book, although many have been mentioned throughout the text. The decision as to how to correct these imbalances will vary in different situations, but in general it is easier and more accurate to correct specific deficiencies for the horses through the feed, and feed the grass for optimum growth as necessary.

Teeth

A grazing horse cannot thrive if it has trouble with its teeth, and may even become dangerous (rearing/head shaking). Tooth problems may manif-

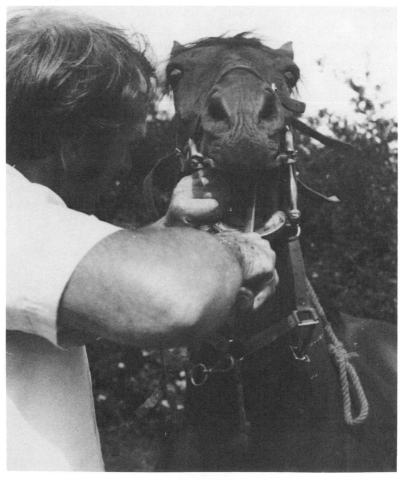

Plate 50 Teeth should be checked and rasped (floated) if necessary at least once a year. (*Photograph by kind permission of Mr Colin Hill.*)

est themselves in a number of ways, including 'quidding' (dropping food from the mouth).

Because of uneven wear on the molars, sharp edges can develop and cut into the tongue or cheeks, leading to soreness and even abscesses. Performance horses should have their teeth checked twice a year and others once a year by your vet or specialist horse dentist. Be sure to warn your vet so he brings a dental gag with him, as attempts at equine dentistry without the proper equipment often 'end in tears' – with a badly upset horse, possibly a bitten tongue, and probably bitten fingers for the vet.

Damage to the mouth may also occur due to grazing short pasture on sandy or gritty soil or to bad bitting, which may affect the ability to eat (e.g. some Argentinian polo ponies arrive in the UK with partially or completely severed tongues due to the vicious bits which their macho riders find it necessary to use as a replacement for good horsemanship).

Very occasionally, horses may lose teeth – e.g. brood mares or youngstock on very calcium or vitamin D deficient diets, and very elderly horses as a result of old age. Check that the diet is balanced, and feed soft or soaked rations which are easily digestible and require minimal chewing. Soaked grass or alfalfa (lucerne) cubes may be a useful supplementary feed for horses kept at pasture who have poor teeth.

So, losing its teeth may not be the 'end of the line' for an old horse, and with a little ingenuity, need not be a great impediment. Very long teeth in elderly horses may again make grazing poor grass, or chewing difficult, so soaked cubes, sugar beet pulp and bread-meal can again be very useful.

Appendix 1
Grass seed mixtures for gallops and racecourses

The primary requirements for the turf on gallops and racecourses are that it be hard-wearing, drought resistant, and provide a cushioning 'bottom' to the sward to reduce concussion injuries. Heading dates may also be considered in that they influence the earliness or lateness of the varieties in question, and it may be possible therefore to extend the working season of the turf. Nutritional value and palatability are obviously not priorities other than that sheep are often used to graze gallops, and even racecourses, as a sward management aid. It should be borne in mind that whereas gallops may be subjected to heavy use daily for up to six months of the year, some racecourses will only be subjected to odd days of extremely hard wear with longer periods for rest and recovery in between.

Seed mixtures for *gallops* are likely to include up to 30% of turf forming pasture perennial ryegrasses, (e.g. Melle and Angela), about 15% creeping red fescue, 20% of a 'hard' fescue such as 'chewings' fescue, 10% hard-wearing Highland bent grass (Agrostis), plus 25% smooth- or rough-stalked meadow grass depending on soil type; seed rate would be approximately 33–37 kg/ha (30–33 lb/acre).

For *racecourses*, the pasture perennial ryegrass levels might be increased to 50% (e.g. 30% Angela and 20% Melle), 20% smooth- or rough-stalked meadow grass (depending on soil type), 20% creeping red fescue, and 10% timothy (e.g. a mixture of 5% Aberystwyth S50 and 5% Aberystwyth S48). Seed rate would be approximately 50–70 kg/ha (45–62 lb/acre).

Specific sward management of gallops and racecourses is beyond the scope of this book. However it is a good idea to check drainage and irrigation requirements. If a new facility is being established I strongly suggest that a full soil survey of the proposed location be carried out beforehand. An accurate soil survey can also be used to predict the going on gallops and racecourses on a day-to-day basis. The addresses of The Soil Survey for England and Wales, and for Scotland are given at the end of this book. Some independent consultants who are members of BIAC can also perform soil surveys overseas. See address list.

Glossary

As fed – Concentrates or forage in the form they would be fed to the animal. *Dry matter* content is the 'as fed' feed with all the water removed at 100°C, and enables nutritionists to compare 'like with like' the nutrient levels in feed without the variable influence of water content.

Bioassay – Quantitative estimation of biologically active substances by the amount of their actions in standardised conditions on living organisms or parts of organisms.

Caecum (blind gut) – Large comma-shaped sac between small intestine and colon; about 1.25 m long with a capacity of about 25 litres in the adult horse. Site of 'fermentation' of fibrous feeds by digestive micro-organisms.

Chelation – In its nutritional application, chelation is the suspending of a bivalent mineral between two or more amino acids. Effectively a mineral or metal molecule is 'wrapped up' or 'locked up' by the amino acids. It comes from the Greek word 'chel' meaning claw. For example, the iron in haemoglobin is chelated, if it wasn't it would not be able to transport oxygen, as iron oxide (rust) would be formed leading to death of the organism.

Choline – Member of vitamin B complex.

Colloid – A substance present in solution is said to be in the colloidal state when its particles, of size 1–100 μm, are dispersed throughout a continuous medium. They are too small to be seen by the naked eye but can be seen with the electron microscope. Because of their small size, colloids have a very high specific surface, giving them pronounced absorptive properties. Clay and humus are examples of colloidal electrolytes. A cubic foot (35.7 kg) of colloidal clay particles has been estimated to have a total surface area of about 60.7 ha (150 acres). Colloidal clay confers on soils high water-holding capacity, which is why clay soils shrink and crack when they dry out.

Diuretic – Causes increase of output of water by the kidney.

Endotoxins – Toxins which form an integral part of some bacterial cell walls are released on the death of the organism. Involved in a number of digestive disorders including laminitis.

Invertebrates – Members of the animal kingdom without backbones, e.g. protozoans in soil, unlike bacteria which are members of the plant kingdom.

Ion – An electrically charged atom or group of atoms. Positively charged ions (cations) have fewer electrons than is necessary for the atom or group to be electrically neutral; negative ions (anions) have more. Ions in solution are due to the ionisation of the dissolved substance.

Legume – Plant coming from the family *Leguminosae* usually associated with nitrogen-fixing bacteria in root nodules. Includes many trifoliate species (e.g. clover). Many bear their fruit in pods (e.g. peas, beans). Others include sainfoin, lucerne (alfalfa), fenugreek, vetches and trefoils.

Ley – Grass or herbage sown with a specific life-span in view, of one to six years depending on herbage varieties used, i.e. not intended as 'permanent pasture'.

Nitrogen fixation – Conversion of atmospheric nitrogen into organic nitrogen compounds, a process that can be carried out only by certain soil-inhabiting bacteria and certain blue-green algae (*Cyanophyta*). Some nitrogen-fixing bacteria live symbiotically with leguminous plants (clover, vetches, trefoil, lucerne/alfalfa, sainfoin, peas, beans, etc.) in nodules on their roots, others live independently in the soil. By their activity soil is enriched with nitrogen to a degree which is of considerable practical importance.

Osmosis – The flow of water or other solvent through a semi-permeable membrane (e.g. cell walls in grass seeds); i.e. a membrane which will permit the passage of the solvent but not of dissolved substances.

Pan – A layer in the soil profile, usually compacted, which prevents free movement of water in the soil. Very strong, dark coloured pans of iron and humus may also be found on some light heath soils, e.g. in Dorset and Surrey. Cultivations or deep cracking in a dry summer help to break up these pans.

Photosynthesis – The process whereby green plants manufacture their carbohydrates from atmospheric carbon dioxide and water in the presence of sunlight. This highly complex reaction is summarised by the equation

$$6CO_2 + 6H_2O \overset{\text{sunlight}}{\rightleftharpoons} C_6H_{12}O_6 + 6O_2$$

i.e. six molecules of carbon dioxide plus six molecules of water produce one molecule of carbon dioxide and six molecules of oxygen. When light falls upon green plants the greater part of the energy is absorbed by small particles called chloroplasts, which contain a variety of pigments, amongst them compounds called chlorophylls, which transform the energy of the light into chemical energy by a process which is not fully understood, but which is known to involve the photolysis (break-down) of water and the activation of adenosine triphosphate (ATP). This energy-rich ATP subsequently energises the fixation of the CO_2, after a series of reactions, so that sugar molecules (carbohydrate) are formed. As animals are unable to fix atmospheric CO_2 in this way, they depend for their carbon on the plants (or other plant-consuming animals) which they consume. Photosynthesis is therefore essential to all higher forms of life, either directly or indirectly.

Probiotic – Substance, usually viable bacterium, used to promote beneficial digestive micro-organisms in the gut, often by restricting the multiplication of harmful ones (e.g. *E.coli*). Used to stabilise gut environment in stress or disease situations, and to help establish (foals) or re-establish the population of the beneficial micro-organisms after disease (scouring, colic, laminitis, etc.) or stress (travelling, sales, shows, post-operative, changes in feeding/management). Different proprietary brands are of variable efficacy depending on the method of manufacture and organism chosen.

Species (spp.) – The smallest unit of classification of living organisms commonly used; i.e. the group whose members have the greatest mutual resemblance.

Symbiosis – Association of dissimilar organisms to their mutual advantage, e.g. association of nitrogen-fixing bacteria with leguminous plants. The bacteria inhabiting the root nodules manufacture nitrogen compounds from nitrogen of the air which become available to the plant. From the latter, bacteria obtain carbohydrates and other food materials. Other examples are bacteria and protozoans in the hindgut of equines.

'Teart' pastures – Contain high levels of molybdenum which can affect plant and animal health, and lock up copper making it unavailable.

Transpiration – Loss of water vapour by land plants. Occurs mainly from leaves and differs from simple evaporation in that it takes place from living tissue and is therefore influenced by the physiology of the plant.

Trifoliate – Three-leaved, as in clover. Leaf is divided into three separate lobes.

Useful addresses

The Equine Management Consultancy Service
The Old School,
Shrivenham Road,
Highworth,
Wiltshire, SN6 7BZ.
Tel: 0793 764759; 0793 763685

Mark Smallman, Artist
The Stables,
Downswood,
Great Cheverell, Nr. Devizes,
Wiltshire.
Tel: 038081 3304

British Horse Society
British Equestrian Centre,
Stoneleigh, Warwickshire, CV8 2LR.
Tel: 0203 52241

The Donkey Sanctuary
Mrs E. M. Svendsen, M.B.E.
Slade Farm, Sidmouth,
Devon, EX10 2NU.
Tel: 03955 6391

National Foaling Bank,
Meretown Stud,
Newport,
Shropshire, TF10 8BX.
Tel: 0952 811234

British Institute of Agricultural Consultants
84 Wellingborough Road,
Irthlingborough,
Northamptonshire, NN9 5RF.
Tel: 0933 650615

Ministry of Agriculture, Fisheries and Food/ADAS
(see your local telephone directory)

Soil Survey of England and Wales
Rothampstead Experimental Station,
Harpenden,
Hertfordshire, AL5 2JQ.
Tel: 058271 63133

Soil Survey of Scotland
Macauley Institute for Soil Research,
Craigiebuckler,
Aberdeen, AB9 2QJ.
Tel: 0224 38611

Wessex Soil and Land
Mr Steve Staines,
2 Nightingale Court,
East Street,
Blandford,
Dorset, DT11 7ED.
Tel: 0258 53695

Wayland Drainage
Mr Tony Clark,
Home Farm,
Shrivenham,
Swindon,
Wiltshire.
Tel: 0793 782235

Water-diviners:
Pat Lucas
Water Diviner
Folly Cottage,
Whitebrook, Nr Monmouth,
Gwent, NP5 4TY.
Tel: 0600 860482

Society of Dowsers
Sycamore Cottage,
Hastingleigh, Nr Ashford,
Kent, TN25 5HW.
Tel: 0233 75253

Henry Doubleday Research Association (re comfrey)
20 Convent Lane,
Bocking,
Braintree,
Essex.

NCMH
c/o City and Guilds,
46 Britannia Street,
London WC1X 9RG.

Agricultural Training Board
Bourne House,
Beckenham,
Kent.
Tel: 01 650 4890

Duffield ATV
Mr Duncan Grove
Brunel Road,
Churchfields,
Salisbury,
Wiltshire, SP2 7PU.
Tel: 0722 334369

Seed houses:
Hunters of Chester Ltd
8 Canal Street,
Chester,
Cheshire, CH1 4EG.
Tel: 0244 47574

Miln Marsters Group Ltd
Waterloo House,
Waterloo Street,
Kings Lynn,
Norfolk, PE10 1PA.
Tel: 0553 3911

Hurst/Finney Lock Seed Ltd
Avenue Road,
Witham,
Essex, CM8 2DX.
Tel: 0376 516600

Equine Behaviour Study Circle
(Chairman:) Hon. Mrs Moyra Williams PhD,
Leyland Farm,
Gawcott,
Buckingham.
Tel: 0280 813301

Auburn University Fescue Toxicity Diagnostic Centre,
Department of Botany, Plant Pathology and Microbiology,
Auburn University,
Al 386849,
USA

Selected useful references and further reading

ISBN – Order Numbers are given where possible.
Dates refer to my copies, more recent ones may be available.

(1) *Breeding Management and Foal Development*, (1982, Equine Research Inc.) PO Box 9001, Tyler, Texas 75711, USA
(2) *Horse Breeding and Stud Management*, Henry Wynmalen M.F.H. (1950).
(3) *The Practical Stud Groom*, (1930) Harry Sharpe.
(4) *Horse Feeding and Nutrition*, Gillian McCarthy. David & Charles (1987) ISBN 0–7153–8672–7.
(5) *The Professional Handbook of the Donkey*, Elisabeth D. Svendsen *ed*, published by The Donkey Sanctuary (1986).
(6) *Horse and Stable Management*, Jeremy Houghton Brown and Vincent Powell-Smith. Collins (1984), ISBN 0–00–3833169–8.
(7) *Grassland Management*, John Parry and Bill Butterworth. Northwood Publications (1981), ISBN 7198–2518–0.
(8) *Grass Farming*, 4th edn M. McG. Cooper and D. W. Morris, Farming Press Ltd, (1977) ISBN 0–852–36076–2.
(9) *Agricultural Botany*, N. T. Gill and K. C. Vear. ISBN 7156–0004–4.
(10) *Lime and Liming*, HMSO (35) ISBN 0–11–240639–4.
(11) *Hunter's Guide to Grasses, Clovers and Weeds*, P. J. P. Hunter.
(12) *British Poisonous Plants*, HMSO Book 161, ISBN 0–11–240461–8.
(13) *Weed Control in Grassland, Herbage Legumes and Grass Seed Crops*, Booklet 2056 (get the latest amendment). Published by the Ministry of Agriculture, Fisheries and Food.
(14) *Comfrey, Past, Present and Future*, Lawrence D. Hills Faber, (1976) ISBN 0–571–11007–X.
(15) *Comfrey Report*, Lawrence D. Hills, published by the Henry Doubleday Research Association.
(16) *Crop Production Equipment – A Practical Guide for Farmers, Operators and Trainees*, H. T. Lovegrove, Hutchinson Educational Ltd, (1968) ISBN 0–09–085390–3.

(17) *Farm Machinery*, 11th edn. Claude Culpin, Collins, (1986) ISBN 0–00–383173–6.

(18) *Horse-Drawn Farm Implements Part II – Preparing the Soil*, Compiled by John Thompson (1979) ISBN 0–950–57757–X.

(19) *Irrigation – Why and When*, Ministry of Agriculture, Fisheries and Food, booklet 2067 (revised 1984).

(20) *Daily Calculation of Irrigation Need – The Water Balance Sheet Method*, Ministry of Agriculture, Fisheries and Food, booklet 2396 (1984).

(21) *Horse Ailments and Health Care*: Colin Vogel (includes a feeding section by Gillian McCarthy) Ward Lock (1982) ISBN 0–7063–6199–7.

(22) *Strongylus Vulgaris in the Horse: its Biology and Veterinary Importance*, C. P. Ogbourne and J. L. Duncan. Available from the Commonwealth Agricultural Bureaux.

(23) *Farming and the Countryside*, (a free publication with useful notes about conservation) MAFF booklet 2384.

(24) *Nature Conservation for Busy Farmers*. Ruth and Andrew Tittensor. Published by Ruth and Andrew Tittensor Consultancy, Walberton Green House, The Street, Walburton, Arundel, Sussex BN18 09B.

There are numerous excellent books and booklets about grass, herb and weed identification.

Index

Page numbers in **bold** indicate references to illustrations.